Manual of
Built-up Roof Systems

THE AMERICAN INSTITUTE OF ARCHITECTS

BY

C. W. Griffin, Jr., P.E.
For The American Institute of Architects

Manual of
Built-up Roof Systems

M c G R A W - H I L L   B O O K   C O M P A N Y

New York
St. Louis
San Francisco
Dusseldorf
London
Mexico
Panama
Sydney
Toronto

MANUAL OF BUILT-UP ROOF SYSTEMS

07-001489-2

890 KPKP 798

Foreword

This manual has been written with the financial support and technical assistance of roofing industry trade associations, manufacturers, and contracting firms. The manual is not, however, designed to serve as an advertising medium for any product or system.

The project was the result of a conference growing out of AIA's concern over built-up roofing system problems. With the assistance of the financial contributors, an Editorial Advisory Board was selected to act independently in the realm of technical discussions and recommendations.

The American Institute of Architects provided a forum for expression of industry opinion and functioned like a university research group. Project Director for the AIA was Frank L. Codella, Administrator, Department of Professional Services.

> Rex Whitaker Allen, FAIA
> *President, The American Institute of Architects*

Acknowledgments

The American Institute of Architects gratefully acknowledges the financial contributions of the following firms and trade associations, which contributed their time and expertise as well in the compilation of this manual.

Allied Chemical Corporation
The Asphalt Institute
Asphalt Roofing Industry Bureau
Carolinas Roofing and Sheet Metal Contractors Association
Certain-teed Products Corporation
The Dow Chemical Company
The Flintkote Company
B. F. Goodrich Company
Insulation Board Institute
Johns-Manville Products Corporation
Koppers Company Incorporate
Lexsuco Incorporated
Lloyd A. Fry Roofing Company
National Roofing Contractors Association
Owens-Corning Fiberglas Corporation
Perlite Institute Incorporated
Philip Carey Manufacturing Company
Ruberoid Company
Southwest Petroleum Corporation
Steel Deck Institute

Additional financial support was also contributed by Marble Products Company, Midwest Roofing Company, and Midwest Roofing Contractors Association, Inc.

Preface

The purpose of this manual is to present the latest information on the design and construction of built-up roofing systems in convenient reference form. At present this vital information is scattered through hundreds of technical periodicals, research reports, product information bulletins, and textbooks.

As a secondary goal the manual will, we hope, stimulate research in improved application techniques, product development and performance, and ultimately, an improved organization of this vital, yet still primitive and badly split, segment of the building industry.

The manual is not an encyclopedia of technical information on built-up roofing systems. It aims merely at highlighting the information needed by the roof designer in his daily work, giving him a basic understanding of the technical and even financial problems encountered in roof design and directing him to the proper sources of information required for this task.

The primary audience is practicing architects who may confront a wide range of troublesome roofing problems almost daily. A secondary, possibly even larger audience, comprises design and maintenance engineers, building contractors, roofing contractors, material producers, and owners concerned with the technical problems of long-range maintenance.

The scope of this manual is limited to the design, specification, installation, and inspection of built-up roofing systems. It focuses on conventional built-up, multiply, felt-and-bituminous roofs, which still predominate. Less intensively covered are the newer roofing materials—

elastomeric or synthetic sealants applied as fluids or in prefabricated sheets, used independently or in conjunction with conventional bituminous materials. A future edition will cover these materials more intensively when more authoritative data are available. Omitted are roof systems surfaced with overlapping shingles, tiles, or sheet metal.

The information presented is the best currently available from a wide spectrum of sources—laboratory researchers, material manufacturers, practicing architects, engineering consultants, roofing contractors, and field foremen.

I am grateful for the technical assistance and editorial criticism of many design professionals and other roofing industry experts. Notable among these are the members of the AIA Editorial Advisory Board for this project: Chairman R. Lloyd Snedaker, FAIA, partner Snedaker, Budd & Watts; William C. Cullen, chief of the Materials Durability and Analysis Section, Building Research Division, National Bureau of Standards; Stephen A. Kliment, AIA, associate partner, Caudill, Rowlett & Scott; Ernest L. Ostic, chairman of the Asphalt Roofing Manufacturers Association Committee on Built-up Roofing; Walter K. Platt, engineer, American Telephone & Telegraph Company; and Edward T. Schreiber, President of Construction Consultants, Inc.

Others who made especial noteworthy criticisms or comments include Paul L. Morris of the Midwest Roofing Contractors Association; Sidney H. Greenfeld of the National Bureau of Standards; and Karl Potasnik of Construction Consultants, Inc.

Unlike the structural design of an isotropic material like steel, or mechanical or electrical design, roofing design is still a primitive and complex art, struggling to become a science. The anisotropic nature of the many roofing materials, whose behavior is further complicated by countless possible combinations and the difficult, sensitive problems of field control, promote uncertainty and complexity. Roofing experts disagree on many vital questions; on some important subjects there is not even a majority consensus.

In resolving controversies, I have applied the following rules: On subjects with an identifiable majority consensus among the consulted experts—notably vapor barriers—I have presented the contradictory viewpoints and followed the recommendations of the majority. On subjects lacking a majority consensus, I have merely presented the contradictory viewpoints along with their explanations. Since an unresolved technical question gives the reader little guidance, the goal is always to resolve these questions, and here I owe a special debt of gratitude to William C. Cullen, who served as umpire.

C. W. GRIFFIN. JR.

Contents

1. Introduction 1

2. The Roof as a System 7

ix

12. *New Roofing Membranes* *192*

ONE

Introduction

The volume of built-up roofing annually installed in the United States totals 2 billion sq ft, enough to cover Washington D.C. and part of its environs. Probably 10 to 15 percent of the roofs included in this vast area fail prematurely. In severely cold regions, the incidence of roof problems exceeds the national average. Even in mild southern California, more than half of 163 surveyed buildings, two to fifteen years old, had a history of leaking roofs. And of 1,000 bonded built-up roofs investigated in another survey, one-third were in trouble within a year or so of their completion.

These roof failures are causing growing concern within the roofing industry for several reasons. A roof failure is expensive; reroofing normally costs twice as much as the original installation. Today's more sophisticated building owners are more conscious of roofing and costs than their predecessors were. Architects and other members of the design professions, threatened by a rising tide of malpractice suits, must devote more attention to roof performance and the soundness of their designs.

WHY ROOFS FAIL

A number of factors contribute to the high incidence of roofing failures. Basically, they spring from the field-manufacturing process, which makes the built-up roofing system one of the most complex field-assembled subsystems of a building. As new products—decks, vapor barriers, insulation, adhesives, and flashing materials—appear on the market, the task becomes more complex.

To the wood sheathing and cast-in-place concrete roof decks that formerly predominated have been added newer deck materials—poured-in-place and precast gypsum, precast concrete, preformed wood-fiber planks, lightgage steel and aluminum, and asbestos cement. The growing use of insulation, required to reduce airconditioning costs, heightens the risk of membrane splitting and condensation within the built-up roof system. The threat of condensation, in turn, creates the possible need for another roof component—a vapor barrier designed to intercept the flow of water vapor into the insulation, where it can cause a host of troubles—membrane wrinkling, blistering, leakage, or destruction of the insulation itself. The vapor barrier may create the need for venting the insulation. And so it goes, with each solution creating its own subproblems.

Thus the roof designer must never consider a component in isolation; he must always investigate its compatibility with other materials and its effects on the whole system. Far more important than the quality of the individual materials are their design and installation as compatible components of an integrated system.

Expanding roof plan dimensions are a source of roofing troubles. Unlike the old multistory loft buildings, where the walls constitute the major exposed area of the building, a modern single-story industrial building, 400 × 1,000 ft in plan and 16 ft high, has nearly 10 times as much roof area as wall area. The trend toward huge, sprawling, single-story industrial buildings follows the evolution of modern assembly-line processes, in which raw materials enter at one end of a building and emerge fully fabricated at the other end, possibly a quarter of a mile away.

Its greater size alone makes a large roof a more complex technical problem than a small roof. A roof over 300 ft long must have one or more expansion joints to accommodate thermal expansion and contraction. On the other hand, a roof only 100 ft long may require no expansion joint. A large roof with a vapor barrier is more vulnerable to blistering than a similar, smaller roof. Blisters in a roof membrane indicate entrapped air and moisture heated into a high-pressure gas that must be relieved by venting to the atmosphere. For a given roof plan shape, the volume of air and water vapor to be released per foot of roof perimeter doubles as the perimeter doubles (since the area quadruples). Thus, on a large roof, the internal pressure producing the blisters is greater than it is on a small roof, and the problem of venting becomes more critical.

Large, level built-up roofs also run a greater risk of inadequate drainage—a major cause of roofing failure. A ponded roof runs the risk of membrane delamination, resulting from freezing water that has penetrated into the plies. Fungi growth, promoted by standing water, can deteriorate organic roofing materials. Irregular ponding can create a warping pattern of surface elongation and contraction, wrinkling the membrane. Yet the lure of first-cost economy, achieved through the simpler fabrication of level steel framing or the elimination of sloped concrete fill, often seduces owners into accepting a dead-level roof.

The effort to economize in roof construction often exerts a strong temptation for some designers. Since the roof is normally a purely utilitarian part of the building, the designer is willing to lavish design effort and client money on more visible parts of a building to the detriment of the roof. "Nothing is expected to do so much, for so many, for so little as the built-up roofing membrane," says one manufacturer's representative.

FIG. 1-1 *The scene above is not an irrigated wheat field; it is a level built-up roof system with ponded water. The roots of the flourishing vegetation can force their way through the membrane into the insulation, ultimately producing widespread leaks. (Sellers & Marquis Roofing Co.)*

FIELD TROUBLE

Despite faulty design, poor field work ranks as the chief cause of roofing failures.   According to Werner H. Gumpertz, a prominent roofing consultant, poor workmanship accounts for the majority of roofing failures. Design errors, structural framing deficiencies, and material failure account for the remainder, in that order.

Faulty field practices spring from two root causes:

- The entry of inexperienced, incompetent roofing contractors into the roofing business
- Accelerated construction schedules

Under the threat of liquidated damages extracted by the owner if the project is not completed on schedule, the general contractor often forces his roofing subcontractor to install the roof before the deck is ready, or in damp, rainy, or severely cold weather.   Poured decks (concrete or gypsum) require a longer time to dry out than other deck materials.   The premature mopping of bitumen can produce a crop of blisters, formed when the trapped moisture evaporates in the spring or summer heat. The water vapor expands to 1,000 or even 1,500 times its liquid volume

FIG. 1-2 *Overweight mopping of surface asphalt produced this classic case of alligatoring (shrinkage cracking resulting from continued oxidation, aging, and embrittlement.).   (Flintkote Co.)*

FIG. 1-3 *Researcher Thomas H. Boone of the National Bureau of Standards, Building Research Section, studies the performance of different membrane samples under severe temperature and moisture cycling.*

inside the membrane plies or at the interface between insulation and membrane.

NEED FOR RESEARCH AND STANDARDS

The roofing industry has lagged in the promulgation of installation standards and test methods, focusing on component material quality instead of the whole field-manufactured roofing system. There are appropriate American Society for Testing and Materials (ASTM) and federal standards for testing important properties of surfacing aggregates, felts, bitumens, insulation, vapor barriers, and structural decks. But there are no generally accepted tests for performance of the entire built-up roof system assembled from these components.

There is no general consensus among the major segments of the roofing industry even on specific field practices. Manufacturers disagree with roofers on a number of vital points, notably:

- The acceptability of "phased" application, in which the lower plies of a built-up membrane are applied and left exposed from a day or two to several months before the upper plies and surfacing are installed
- The tolerance for bitumen weights in interply moppings
- Kettle temperature tolerances for heating bitumen
- Ambient temperatures at which hot bitumen can be applied

Building code officials lag in adapting their requirements to the increasing complexity of contemporary roof design.   Few standard specifications recognize ponding as a problem requiring special deflection limits for the lengthening spans in contemporary roof framing.   The Open-web Joist Specifications, High-strength Series, for example, limits roof live-load deflection to $\frac{1}{360}$ of the span for members supporting a plastered ceiling, and $\frac{1}{240}$ of the span for all other cases.   The permitted 5-in. deflection on a 100-ft span joist could produce unstable, progressive deflection under rainwater load, resulting in roof ponding and even in collapse (see Chap. 3, "Structural Deck").

Much more research is needed on wind-uplift resistance, a matter of growing concern to insurance companies.   Many popular materials remain untested for wind-uplift resistance.   Laboratory tests of various insulation boards, notably for resistance to delamination, are needed. Cold-applied adhesives and mechanical fasteners require broader, more thorough testing to provide a solid basis for wind-uplift design.

Yet despite the many industry problems—ranging from technical controversies to disagreements over insuring policies—there is nonetheless a broad expert consensus on good roofing practices.   This manual attempts to extract this broad consensus from the technical literature and from accumulated expert testimony.

The Roof as a System

Like curtain walls, structural framing, airconditioning, and other building subsystems, the built-up roofing system is an assembly of interacting components designed for a specific function—in this instance, the protection of the building interior, its contents, and its occupants from the weather. Of the built-up roofing system's four basic components (structural deck, vapor barrier, insulation, and built-up membrane), only the structural deck and the built-up membrane are indispensable. As a third component, most modern roofs contain a layer of thermal insulation, normally sandwiched between the deck and the membrane. The less common fourth component is a vapor barrier, placed between deck and insulation. (Each component is discussed in greater technical detail in succeeding chapters.)

In mathematical terms, a systems-designed component is one of several mutually dependent variables rather than an independent variable designed in isolation. Yet each component still plays its own unique role. The structural deck resists dead, live, wind, and possibly earthquake loads. The semiflexible, waterproof membrane, a lamination of felt alternated with layers of bituminous

7

mopping, weatherproofs the roof assembly.   Thermal insulation reduces heating and cooling loads by impeding the transfer of heat through the roof.   It also prevents destructive condensation on interior surfaces and stabilizes the temperature of the structural deck.   The vapor barrier, an essentially impermeable membrane, retards the flow of water vapor into the insulation, where condensation can harm insulation performance or gaseous expansion can accelerate deterioration of the membrane.

Flashing, though not a basic component of the built-up roof system, is an indispensable accessory for sealing joints at gravel stops, walls, expansion joints, vents, drains, and wherever else the membrane is interrupted or terminated.

The built-up roofing assembly (including flashing), functions as a system in which each component depends on the satisfactory performance of the other components.   The integrity of the waterproof membrane depends on good anchorage to the substrate and adequate shear strength between the deck and the vapor barrier, between the vapor barrier and the insulation, and between the insulation and the membrane.   Thermal resistance of the insulation, which can be destroyed by moisture, depends on the effectiveness of vapor barrier and membrane.   It may also depend on the dispersion of entrapped water vapor through edge and stack vents.   The integrity of vapor barrier, insulation, and membrane depends on the stability of the structural deck.

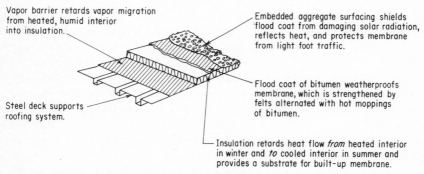

Vapor barrier retards vapor migration from heated, humid interior into insulation.

Embedded aggregate surfacing shields flood coat from damaging solar radiation, reflects heat, and protects membrane from light foot traffic.

Steel deck supports roofing system.

Flood coat of bitumen weatherproofs membrane, which is strengthened by felts alternated with hot moppings of bitumen.

Insulation retards heat flow *from* heated interior in winter and *to* cooled interior in summer and provides a substrate for built-up membrane.

FIG. 2-1   *Roof system components and functions.*

Thermal insulation has a complex effect on built-up roof systems. Largely because it retards heat flow to and from the roof surface, thermal insulation raises the extreme range of roof surface temperatures by up to 50°F (40°F hotter in summer than an uninsulated roof surface, 10°F colder in winter), thus accelerating the photo-oxidative chemical reactions that embrittle surface bitumen and make the membrane more

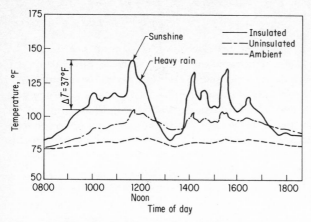

FIG. 2-2 *Extreme roof temperatures produced by insulation sandwiched between deck and built-up roofing membrane can make an insulated roof surface 40°F hotter in sunlight, 10°F colder at night, than an uninsulated roof surface. This extreme temperature cycling accelerates the membrane's deterioration. (National Bureau of Standards Technical Note 231.)*

susceptible to alligatoring and cracking. The greater magnitude and faster rates of temperature change of an insulated roof produce greater thermal stresses and consequent chance of splitting.

An excellent illustration of the interdependence of individual components is the fire test performance of a roof-ceiling assembly. One criterion for qualifying a roof-deck assembly for a given fire rating (in hours) is its resistance to heat flow. As a safeguard against the roof covering igniting, the average surface temperature rise during the furnace test must not exceed 250°F above the initial temperature (see Chap. 8, "Fire Resistance"). Since insulation retards heat flow, the designer might assume that adding more insulation to a rated roof-ceiling assembly must improve its fire performance.

He could be disastrously mistaken. Added insulation would indeed depress the roof surface temperature. But if heat loss through the roof is excessively retarded, it can cause a structural collapse. A lower surface temperature would mean a *higher* ceiling plenum temperature (resulting from undissipated heat). This higher plenum temperature could buckle steel joists or ignite combustible structural members that otherwise would continue to carry their loads. The system designer, like a juggler, must keep his eye on more than one ball. He must never assume that you cannot have too much of a good thing.

STRUCTURAL DECK

Decks are classified as nailable or nonnailable (sometimes both) in describing the method of anchoring the vapor barrier, insulation, or built-up membrane to the deck.   Some decks—notably poured gypsum— can be nailed, hot-mopped with bitumen, or coated with a cold-applied adhesive.   Others, like timber plank or plywood, should be limited to nailing because of the threat of heated bitumen dripping through the joints and drying shrinkage causing membrane splits.   Poured concrete, except in rare instances when wood nailers are cast into its surface, is limited to nonnailed anchorages (see Chap. 3, "Structural Deck").

ANCHORING THE COMPONENTS

The systems approach is well illustrated in the different advantages and handicaps of nailing or mechanically anchoring roof components or in mopping them with hot bitumen or using cold-applied adhesives.   Designers must always consider several factors in any phase of roof design.

Bonding of components for good anchorage is the basic consideration. For suitable surfaces (initially dry and remaining so), solid bituminous mopping is the best general method for anchoring one component to another.   It generally provides the highest horizontal shearing strength and resistance to wind uplift, as well as the best protection against moisture infiltration.

Secondary factors, however, often weigh against solid mopping.   For substrates subject to shrinkage cracking (e.g., poured gypsum), solid mopping increases the hazard of splitting because it concentrates mem-

FIG. 2-3   *Solid anchorage of insulation to the deck is essential to resist wind uplift, thermal and moisture-induced expansion, and contraction stresses.   (Lexsuco, Inc.)*

brane stresses over the cracks. Spot or strip mopping reduces this hazard, as well as providing a lateral avenue of moisture escape and pressure relief.

Nailing, or some other form of mechanical anchorage, also has its special advantages. To prevent felt slippage, mechanical anchorage is required even on hot-mopped decks sloped more than $\frac{1}{2}$ in. per ft (for coal-tar pitch or dead-level asphalt) or more than $1\frac{1}{2}$ in. per ft for steep asphalt. A nailed base felt also provides horizontal passages for escape of water vapor entrapped between the deck and the component above it.

Fire requirements pose a major obstacle to the use of hot bituminous mopping on the widely used steel deck. The use of hot-mopped bitumen as a vapor seal or to bond organic insulation or membrane disqualifies a steel deck for a Class I rating (which eliminates the expense of ceiling sprinklers). For such decks the designer must specify approved mechanical anchors or cold-applied adhesives (see Chap. 8, "Fire Resistance").

DRAINING THE ROOF

Slope and rigidity are the critical factors in preventing accidental ponding of water and ensuring positive drainage of the roof's surface. For low-slope roofs, the normally recommended live-load deflection limits, ranging from $\frac{1}{180}$ to $\frac{1}{360}$ of the span, may be totally inadequate to prevent progressive roof ponding, which in extreme cases can cause a collapse.

Minimum roof slope should be $\frac{1}{4}$ in. per ft. The inevitable ponding on dead-level roofs may freeze and delaminate or split the membrane; it may promote the growth of vegetation, whose roots may penetrate the membrane; and it may produce a warping pattern of surface thermal deformation, thereby wrinkling the membrane (see Chap. 3, "Structural Deck").

VAPOR BARRIERS

Vapor barriers are essentially impermeable membranes, normally placed on the structural deck (below the insulation) to control the flow of water vapor from heated, humidified interiors upward into the insulation or built-up roofing. Under varying temperatures (as liquid, gas, or ice), entrapped moisture can destroy the thermal-resisting value of the insulation; blister, split, or delaminate the membrane; or leak into the building.

Vapor barriers designed to prevent this infiltration cover a wide range of materials. A common vapor barrier, often known as a vapor seal, comprises three bituminous moppings with two plies of felt or two bituminous moppings enclosing an asphalt-coated base sheet. Other vapor-barrier materials include various plastic sheets, glass, aluminum foil, and laminated kraft paper sheets with bitumen sandwich filler, or bitumen-coated kraft paper.

Like insulation, a vapor barrier can cause problems, but unlike insulation, the problems may outweigh the benefits. If the insulation contains moisture when installed, the vapor barrier will help prevent its escape, for it forms the bottom face of a sandwich whose top face is the membrane. Moreover, under the many field threats to its integrity—punctures, cutting of new roof openings, etc.—a vapor barrier almost always admits some water vapor. Thus the prudent designer specifying a vapor barrier will consider a system that permits the escape of water vapor: edge venting, and, on large areas, stack venting to facilitate escape of moisture from the roof sandwich and to reduce the threat of vapor pressure buildup. The foremost roofing experts today reject the traditional advice, *"If in doubt, use a vapor barrier"* in favor of *"If in doubt, omit the vapor barrier."*

THERMAL INSULATION

As its primary functions, thermal insulation cuts heating and cooling costs, increases interior comfort, and prevents condensation on interior surfaces. Its secondary functions are almost as important. Through its horizontal shearing resistance, insulation helps to relieve concentrated stresses transferred to the built-up membrane from movement in the structural deck. It also provides an improved substrate for application of the membrane on a steel deck.

Insulation comes in many materials—rigid insulation prefabricated into boards, poured insulating concrete fills (sometimes topped with another, more efficient rigid board insulation), or dual-purpose structural deck and insulating plank. Chemically, insulations are *organic*—the various vegetable-fiber boards—or *inorganic*—the glass fiber, foamed glass, perlite board, or insulating concretes with lightweight mineral aggregates.

Fiberboard insulations are generally most vulnerable to moisture, which will eventually rot and weaken any fibrous organic vegetable board or organic plastic binder. But all insulation materials are vulnerable to moisture or freeze-thaw damage (see Fig. 2-4).

Insulation must have the following structural properties:

- Good shearing strength, to distribute tensile stresses in the membrane and prevent splitting
- Compressive strength to withstand traffic loads and (especially in the Midwestern states) hailstone impact
- Adhesive and cohesive strength to resist delamination under wind uplift
- Dimensional stability under thermal and moisture changes

FIG. 2-4   *Moisture trapped within the sandwiched insulation is a major threat to built-up roofing systems.   If the surface moisture on the closed-cell insulation is not allowed to dry before application of the membrane (top), it may blister the membrane when the moisture evaporates and expands under solar heat (bottom).   Membrane cracking and leaks normally follow blistering long before the expected 15- or 20-year life expectancy of such a roof has expired.   (GAF Corporation.)*

Under this host of demanding requirements, the design of thermal insulation and choice of materials become one of the roof designer's most complex tasks.

BUILT-UP MEMBRANE

The built-up membrane, the weatherproofing component of the roof system, has three basic elements: *felts* and *bitumen*, alternated like a multideck sandwich, and a *surfacing*, normally of mineral aggregate. It forms a semiflexible roof covering, with as few as two or as many as five plies of felt, custom-built to fit the contours of the deck (see Fig. 2-5).

The bitumen—coal-tar pitch or asphalt—is the waterproofing agent.

The felts form a fabric matrix that stabilizes and strengthens the bitumen. They prevent excessive flow when the bitumen is warm, and, when the bitumen is cold, tend to distribute stresses over large areas. Felt reinforcement is either *organic*—generally manufactured from felted papers or shredded wood fibers—or *inorganic*—asbestos or glass fiber.

Mineral aggregate surfacing, normally gravel, crushed rock, or blast-furnace slag, protects the bitumen from life-shortening solar radiation. Because of its damming action, aggregate permits the use of heavy, uniform pourings of bitumen (up to 75 lb per square), with consequent better waterproofing and longer membrane life. It also serves as a fire-resistive skin, preventing flame spread and protecting the bitumen from the abrasion by wind, rain, and occasional foot traffic.

On "smooth-surfaced" asphalt roofs with asbestos or glass-fiber felts or, less frequently, organic felts, surfacing aggregate is sometimes omitted.

FIG. 2-5   *The normal roofing practice of shingling felts (three plies over a single-ply coated base sheet in the four-ply built-up roofing membrane shown above) prevents slippage between felts and interply moppings. Shingling also facilitates field application.   (Flintkote Co.)*

The chief advantages of smooth-surfaced roof are (1) reduction in roof dead weight (by 3 or 4 psf) and (2) less difficulty in locating and repairing leaks obscured by the aggregate. Smooth-surfaced roofs are, however, less durable, and they require more frequent maintenance than aggregate-surfaced roofs.

Design of the built-up membrane also requires a knowledge of what the membrane cannot do, as well as what it can and must do. No membrane has the strength to resist large movements in the deck, insulation, or other components in the built-up roofing assembly.

Built-up membranes cannot resist puncture by sharp objects. Wherever more than occasional light foot traffic is expected on roofs, and especially where workmen may drop steel tools, the designer should provide walkways.

NEW ROOFING MEMBRANES

Since 1957, elastomeric roofing membranes have appeared on the market, chiefly as a simpler way of covering hyperbolic paraboloids, barrel shells, folded plates, and other irregular roof surfaces. Along with cold-applied bituminous membranes, these new coverings also mark a continuing effort to control the cost of field labor. These new labor-saving roof coverings are either fluids—sprayed, rolled, or brushed—or sheets—prefabricated for direct field application, with conventional or special adhesives. Experience with these coverings, however, remains too limited for authoritative recommendations.

FLASHING

Flashings are classified either as *base flashings*, which form the upturned edges of the membrane where it is pierced or terminated, or *cap flashings*, which shield the exposed joints of the base flashing from above. Base flashings are normally made of bitumen-impregnated felts or fabrics, plastics, or other nonmetallic material. Cap flashings are often made of sheet metal, including copper, lead, aluminum, stainless steel, or galvanized steel.

As the major source of roofing leaks, flashing demands as much of the designer's attention as the basic built-up roofing components.

DESIGNING THE ROOF SYSTEM

The best roof design for any project represents a synthesis of many factors. The most basic requirement is to satisfy the building code and

FIG. 2-6   *The support of the expanding-contracting pipe*
(shown above), a loose assembly of insulation boards and
timber blocking, has slipped, exposing the bitumen flood coat
to weather deterioration.   The designer should have specified
flashed pedestals carried on the roof's structural framing,
with roller bearings for the pipe.

probably the insurance company requirements for fire and wind resistance.   Major design factors are:

- First cost and ultimate cost
- Value and vulnerability of building contents
- Required roof life
- Type of roof deck
- Climate
- Ease of maintenance
- Availability of materials and component applicators
- Local practices

Design of the built-up roof system requires consultation with other
members of the design team.   The architect must confer with the
mechanical engineer on heating and cooling loads to design the insulation
and to keep roof penetrations to a minimum, thereby reducing the chances
for flashing leaks.   He must work with the structural engineer to assure
slope and framing stiffness that will avoid ponding.   And he must advise
the owner to institute a periodic maintenance and inspection program to
avoid clogged drains, which may pond rainwater to excessive depths and
possibly collapse the roof.

CONSTRUCTING THE ROOF

Under the present organization on a normal roofing project, the work ideally proceeds as follows:

- The architect specifies the roofing components and installation procedures and submits progress and final inspection reports before acceptance.
- The roofing manufacturer furnishes products complying with the specifications and cooperates with the roofer to ensure that materials are dry when delivered to the site. Under the normal manufacturer's bonded roof program, the manufacturer inspects the work of the roofing subcontractor, who has previously demonstrated his ability to install the manufacturer's materials.
- The general contractor schedules and coordinates the work of the roofing subcontractor and other subcontractors working on the deck (plumbers, electricians, heating, airconditioning, and ventilating workmen, etc.), and makes sure that the stored materials are kept dry. He also provides the roofing subcontractor with a satisfactory deck surface.
- The roofing subcontractor performs the actual field work, coordinated by the general contractor. He installs the vapor barrier, insulation, built-up membrane, and, usually, the flashing.

Along with this apportionment of responsibility, a roofing job may be covered by a roofing manufacturer's *bond*. A roofing bond, issued by the manufacturer and backed by a surety company, guarantees the owner that the surety company stands behind the manufacturer's liability to finance membrane repairs required to stop leaks occasioned by ordinary wear. Most bonds are written for limited liability, generally for a period of 20 years. (For more detailed discussion of roofing bonds, see Chap. 10, "Specifications and Performance Criteria.")

DIVIDED RESPONSIBILITY

Preceding a roofing failure on a typical job, where the manufacturer's bond may lull the architect and owner into a false sense of security because they have not studied the bond provisions, may lie the following sequence of events:

- The designer relies on the integrity of the prequalified roofer and the manufacturer and skimps on roofing system details.
- The roof subcontractor exercises his option to select a cheaper roof specification by a manufacturer whose product still qualifies for the specified 20-year bonded roof.

- The general contractor, disregarding the roofer's qualifications and ignoring the application technique himself, selects the low roofing bid and relies on the manufacturer's inspection required under the bond.
- The manufacturer's inspector, often the salesman who sold the materials to the roofer, finds himself charged with inspecting the work of a customer on whose continued good will he depends for future material sales.

Assigning liability for a roof leak on such a job may pose a formidable challenge in determining whether the trouble stems from faulty design, from careless field work by the roofing subcontractor, from defective materials supplied by the manufacturer, or from an uneven structural deck installed by the general contractor or another subcontractor. The technical complexity of modern built-up roofing systems, with proliferating material combinations creating new and sometimes unforeseen interactions between components, puts new stress on the creaking structure of responsibilities. Their diffused apportionment permits, and even encourages, a round or two of buck-passing.

UNIFIED RESPONSIBILITY

On intricate mechanical or electrical systems, a simpler, more rational apportionment of responsibilities has evolved. On an elevator subcontract, for example, the architect sets the general performance standards. The manufacturer advises the architect on hatchway and door clearances, access, power outlets, and other requirements. As the building rises, the elevator company's construction superintendent arranges for installation of the rail brackets or inserts.

The elevator contract is essentially complete when the architect or owner accepts the installation, following the manufacturer's testing program. Under a contract provision, the elevator manufacturer maintains the equipment generally for a 3-month period. The more unified responsibility works better than the splintered responsibility of a typical roofing subcontract.

SYSTEMS DESIGN

One roofing expert has proposed that a comprehensive systems specification, applied to a built-up roof system, would require of the designer:

- A thorough knowledge of how the components fit together
- Access to a well-informed manufacturer's representative and a qualified roofing contractor, who could supply accurate, current technical information
- Approval of all roof details included in the contract drawings
- Inclusion of all pertinent information in the specification

table 2-1   GUIDE TO ROOF SYSTEM PERFORMANCE

(Based on table in "Study of Roof Systems and Constituent Materials and Components," Building Research Advisory Board, SAR 6)

Performance requirement	Deck	Vapor barrier	Insulation	Membrane	How evaluated	Available test adequate? (ASTM test number given, except as noted)	Study feasibility of modifying test?	Need new test?
		Applicability						
Weather resistance	...	...	...	x	[a]	No	...	[b]
Liquid-water resistance	...	x	...	x	Test[c]	Yes (UL or FM)	...	Yes
Wind resistance	x	x	x	x	Test	Yes (E84, E108, E119, E163)	...	Yes
Fire resistance	x	x	x	x	Test		...	
Impact-puncture resistance	x	x	x	x	Test	No	...	Yes
Bitumen flow resistance				x	Test	Yes (D466)	Yes	Yes[d]
Slippage resistance				x	Test	No	Yes	
Flexibility				x	Test	No	...	
Abrasion resistance				x	Test	No	Yes	
Strain resistance	x	x	x	x	Test[e]	Yes (D2523)	Yes	
Thermal-movement resistance		x	x	x	Test	No	...	
Thermal-shock resistance	x	x	x	x	Test	?	Yes	
Water-vapor resistance		x	x	x	Test	Yes (E96)	...	
Adhesion	x	x	x	x	Test	Yes (D903)	Yes	
Fungus-attack resistance	x	x	x	x	Test	No		
Appearance								
1 Permanent deformation					Judgment			
2 Surface defects					Judgment			
3 Nonuniform color					Judgment			
Resistance to noise generation	x	x	x		Judgment			
Thermal insulation			x	x	Test	[f]		

[a] Test currently not available.
[b] Improbable that test can be devised.
[c] Except for sloping deck with continuous covering.
[d] With respect to creep.
[e] Partly a design factor.
[f] Except for determining how thermal-insulating value endures through service life.

The successful formulation of testing techniques to improve roof performance possibly marks the next step in the evolution of roofing systems toward the integrated systems approach.   Ideally, under this approach, the architect sets the general design and performance standards for the roof, and a single subcontractor details, fabricates, erects, tests, and guarantees the roof's performance.   Such a guarantee is currently impracticable because of the many trades working on the roof and the common practice of installing the roof before other trades are finished.   Use of a temporary roof, scheduled for complete removal before the permanent roofing system (from vapor barrier on up) is installed, is at least a partial solution to this problem.

As a practical start in extending the systems approach, the Building Research Advisory Board has defined the service performance requirements for built-up roofing systems for the Federal Housing Administration.   There are existing tests available for evaluating some of these requirements—for example, ASTM E-96 for determining vapor-barrier resistance to water-vapor transmission, or ASTM D-903 for testing adhesion.   For most performance requirements, however, there is no readily available test, and for these, the Building Research Advisory Board researchers recommend either modification of existing tests or development of new tests (see Table 2-1).   (For some performance requirements—notably fire resistance and wind uplift—the roof system is tested as a whole; for others—e.g., fungus-attack resistance, water-vapor transfer resistance—individual components are tested.)

Before the systems approach can evolve beyond its primitive contemporary state, the performance criteria set by architects must be complemented with standard tests for evaluating performance—not of individual components, but of the entire roof system.

Improved roof performance depends less on purely technological progress, manifested in the perennial search for new miracle materials, than on a deeper understanding of the roof as a complex system of interacting components.   The correct design combination of materials and good field application outweigh even material quality as ingredients of a durable, weathertight roof.

THREE

Structural Deck

The structural design of the roof deck and its supporting frame to resist gravity loads, earthquake, and wind forces is largely beyond the scope of this manual.   Several secondary structural considerations, however, directly concern the roof system.

The major considerations in designing the roof deck are:

- Deflection
- Component anchorage
- Dimensional stability
- Fire resistance

Rational deflection limits that prevent the formation of rainwater ponds can vitally affect the drainage design of level roofs.   Anchorage of the component layers is essential to prevent delamination of the roof assembly by wind uplift or horizontal movement.   The roof deck's dimensional stability depends on its coefficient of thermal expansion and, in organic fibrous materials, on the degree of swelling accompanying moisture absorption.   Excessive movement of the substrate can wrinkle or split the built-up membrane.   The deck's dimensional stability becomes an especially acute problem when thermal insulation is either omitted or placed below the deck. And the structural deck must carry its design loading through any fire-resistance tests.

DECK MATERIALS

The basic roof decks used with built-up roofing systems are:

1. Wood sheathing—in sawed lumber or plywood sections
2. Preformed, mineralized wood fiber
3. Gypsum—precast or poured-in-place (see Fig. 3-1)
4. Concrete—precast or poured-in-place
5. Metal deck—lightgage, ribbed steel, or aluminum (see Fig. 3-2)
6. Composite corrugated metal deck plus poured-in-place concrete topping designed for integral structural action with the deck
7. Asbestos cement, in ribbed or cellular sections

Roof decks are classified as *nailable* or *nonnailable*.   In general, wood decks (over ½ in. thick), preformed wood-fiber decks, and gypsum decks, both precast and poured-in-place, are nailable.   Nonnailable decks are metal and concrete.   The chief means of anchoring roof components to nonnailable decks is with hot-mopped bitumen, cold-applied

FIG. 3-1   *Workmen screed a fast-setting, wire-reinforced gypsum deck, poured on gypsum form boards spanning between the flanges of closely spaced bulb tees.   (U.S. Gypsum Co.)*

table 3-1 ANCHORAGE TO STRUCTURAL DECK
   Key: F = Mechanical fasteners[a]
        A = Adhesives[b]

Component to be applied	Deck substrate					
	Wood (sawed, plywood)	Preformed wood fiber	Gypsum (poured, precast)	Metal	Concrete (poured, precast)	Asbestos cement
Vapor barrier						
Felt type..............	F	F	F	A	A	A
Plastic sheet...........	F-A	...	...	F[c]-A		
Metal foil.............	F-A					
Kraft paper...........	F-A	F[c]-A	F-A	A	A	A
Insulation						
Mineral aggregate board	F-A	F-A	F-A[e]	F-A	A	A
Vegetable-fiber board...	F-A	F-A	F-A[e]	F-A	A	A
Glass-fiber board.......	A	A	F-A[e]	F-A	A	A
Glass foam board.......	A	A	F-A[e]	F-A	A	A
Plastic foam...........	A	A	F-A[e]	F-A	A	A
Corkboard.............	A	A	F-A[e]	F-A	A	A
Lightweight concrete...	Poured in place					
Built-up membrane.......	F[d]	F[d]	F[d]		A	A

[a] Mechanical fasteners are nails or special fasteners.
[b] Adhesives are hot bitumen or fire-rated, cold-applied mixtures.
[c] Fastened through insulation.
[d] Base ply (or plies) only, unless slope requires back nailing.
[e] Apply coated sheet on gypsum substrate before installing insulation board.

FIG. 3-2   *Lightgage, ribbed steel deck is by far the most widely used deck in built-up roofing systems.   (Granco Steel Products Co.)*

adhesive, or mechanical fasteners that depend on friction or locking action (see Table 3.1).

DESIGN FACTORS

Draining the Roof

Deliberately ponded roofs, kept perpetually flooded, once enjoyed a minor vogue as a means of restricting the bituminous membrane's surface temperature range and adding insulating value in summer.   Roofing experts today reject this design strategy.   Drain the roof; no accidental ponds should be left after rain.

There are many reasons for this advice:

1. Freezing of ponded water that has penetrated into the plies can delaminate the membrane, literally tearing the roof apart.   Ice formed on a roof surface and keyed into the surface contracts nearly 1 in. in 50 ft with a 12°F temperature drop, creating tensile stresses that may split the roof membrane.   Water freezing over an entire roof can, on expanding, push out parapet walls.

2. Standing water can promote the growth of vegetation and fungi, create breeding places for insects, and produce objectionable odors. Plant roots can puncture the membrane and spread into the insulation, threatening leaks, blisters, wrinkles, and destruction of the insulation.

3. Wide temperature variations in a randomly ponded roof can promote a warping pattern of surface elongation and contraction, possibly wrinkling the membrane.

4. It is impossible to apply a new built-up membrane or to repair a roof with ponds of standing water.

5. Evidence of ponded water on a deck after a rainfall nullifies some manufacturers' roofing bonds.

The quest for economy sometimes prompts designers to ignore this basic advice to drain decks. When a roof is designed for conversion to a floor in a future building expansion, it is normally expedient to design a dead-level roof. This practice avoids the additional cost of sloped framing or sloped fill and the trouble and expense of later removal. Even for permanent roofs, the additional costs for sloping a roof may outweigh the advantages. Moreover, a dead-level roof may allow an architect to carry out an aesthetic design that a sloped roof would spoil. But architects should recognize that ponding is inevitable on dead-level roofs. Owners must realize that periodic inspection and maintenance are more urgent for a dead-level roof than for a properly drained roof.

There are two basic roof drainage designs: (1) the perimeter system, normally with scuppers and downspouts; (2) the interior drainage system. Both require a study of rainfall records to design the gutter and downspout capacities. (Design load can be the maximum 5-min or 1-hr rainfall rates recorded for 5- or 10-year periods. For rainfall data for various cities, see "Architectural Sheet Metal Manual," Sheet Metal & Airconditioning Contractors National Association, 107 Center St., Elgin, Ill., or check weather bureau records.)

In perimeter drainage, water flows down from interior roof regions to the perimeter, where it collects in scuppers or gutters and flows into downspouts or merely drips off the roof. Perimeter drainage is generally the least expensive, but has several disadvantages. The higher central elevations required to assure drainage to the perimeter increase the building cubage, especially for a large roof area. Freezing in the exterior drainage can create ice dams, which block drains and cause overflow. Freezing water can even distort the metal and burst gutters and downspouts.

The interior drainage system requires installation of roof drains in the deck itself. Rainwater flows through interior downspouts or leaders, normally located within a column encasement, and sometimes within the column itself. An interior system requires greater care in sizing and locating drains—especially against setting the drains high. This is a common fault in poured concrete decks, where the preset drains are often used as screed guides. Interior downspouts must be insulated, to protect finished areas from condensation.

Interior drains should be located near midspan of the roof framing, where maximum deflection occurs. Locating drains near columns, where there is little or no deflection, increases the chances of rainwater ponding (see Fig. 3-3).

Roof decks are sloped either by sloping the framing or by applying a fill of varying thickness over a horizontal deck surface. Sloping the

FIG. 3-3   *Drain sump pans, tack-welded in place, are furnished by steel deck manufacturers.   (Granco Steel Products Co.)*

framing is potentially a less troublesome, if more costly method.   Light-weight concrete fill has several disadvantages:

- It heightens the chances of entrapping moisture within the roof sand-wich, especially in ordinary metal decks, which lack the continuous open joints designed for escape of the water vapor in some corrugated decks.
- Its coefficient of thermal expansion generally differs from that of the structural deck.
- On deck surfaces where the fill tapers down to a "feather edge," it may break, and the concrete fragments may become a sliding substrate for the built-up membrane.

Deflection

Deflection limits normally set in building codes, manufacturers' bulletins, and other guides or recommendations have little or no relevance to roof problems.   The typical deflection limits—$\frac{1}{180}$, $\frac{1}{240}$, or even $\frac{1}{360}$ of the span for live, or even total load—may be excessive for long-span roofs, which are especially vulnerable to ponding, and too conservative for short-span decks.   Structural failures, including sudden and total col-lapse, have occurred in some poorly designed level roofs.   Most of these roofs satisfied the normal code design provisions.

Ponding attributable to faulty design can be explained as follows: The deflection curve produced by an accumulating weight of rainwater may form a shallow basin in a roof.   If the outflow does not prevent the capacity of the basin from increasing faster than the influx of rainwater, then the roof is unstable, and a long, continued rainfall is structurally hazardous.   Even if the structural deck withstands the ponding load, the standing water threatens the membrane and the insulation and ulti-mately, if the roof leaks, the building contents.

As a rough safeguard against ponding, for level roofs designed with *less* than the minimum recommended slope of ¼ in. per ft, the *sum* of the deflections of the supporting deck, purlins, girder, or truss under a 1-in. depth of water (5-psf load) should not exceed ½ in.

A proper code provision would specify a minimum stiffness, such as maximum deflection = ½ in. for live load = 5 psf (the weight of 1 in. of water). Thus the deflection would increase at roughly half the rate of the rainwater buildup. If the roof structure doesn't satisfy this requirement, then the designer should present computations substantiating the safety of the slope used.[1]

In computing deflections the designer must consider plastic flow in concrete and wood. Plastic flow (increasing deflection under constant prolonged loading) is inevitable in materials like reinforced or prestressed concrete. Laboratory tests indicate plastic strains ranging from 2½ to 7 times elastic strains in structural concrete. Since these larger extremes seldom occur in practice, the normal assumption is that plastic flow adds an increment of 2½ to 3 times elastic deflection. Thus the total deflection attributable to elastic strain + creep = 3 or 4 × instantaneous elastic deflection.[2]

Wood is similarly subject to long-term plastic deformation. For glued laminated members (including plywood) and seasoned sawed members, residual plastic deformation is about 50 percent of elastic deformation. For unseasoned sawed members, plastic deformation will rise to 100 percent, thus requiring a doubling of computed elastic deflection to approximate longterm deflection.

Preformed structural wood-fiber plank sometimes exhibits excessive, and erratically unpredictable, permanent inelastic deflection. These deflections are not a true creep or plastic flow—attributable solely to load-produced strain. A combination of thermal deformation and moisture absorption, which weakens some cementitious binders used in these materials, causes some preformed structural wood fiber units to sag—up to ½ in. in a 4-ft span. Such excessive deflection threatens the entire roof assembly, increasing the chances of membrane splitting or wrinkling and thus of subsequent leakage.

[1] For a more elaborate analysis of required roof slope and framing stiffness, see the *Timber Construction Manual*, pp. 4–153ff, John Wiley & Sons, Inc., New York, 1966, and Roof Deflection Caused by Rainwater Pools, *Civil Engineering*, October, 1962, p. 58.

[2] For more detailed analysis of the part played in the deflection of concrete structural members by shrinkage and temperature deformation, as well as plastic flow, see the American Concrete Institute Proceedings, vol. 57, pp. 29–50, 1960–1961; also see *Deflections of Horizontal Structural Members*, by W. G. Plewes and G. K. Garden, National Research Council, Department of Building Research, Canadian Building Digest 54.

Deflection and deck movement may also be aggravated by foundation settlement. In a roof designed for the minimum recommended $\frac{1}{4}$-in. slope, foundation settlement is normally insignificant. (A differential settlement of 1 in. in adjacent column footings 30 ft apart would reduce a $\frac{1}{4}$-in. roof slope by only 13 percent.) Roof-deck vibration from seismic forces, traffic, or vibrating machinery may also require attention.

Dimensional Stability

The dimensional stability of a roof deck is determined largely by its coefficient of thermal expansion and its propensity to change dimension with changing moisture content. These factors vary greatly with different materials. Inadequately insulated aluminum decks, with a coefficient of thermal expansion nearly double that of steel (13×10^{-6} versus 6.7×10^{-6} per °F), present an especially serious problem.

Wood has a relatively low thermal coefficient of expansion longitudinally. It is from 1.7×10^{-6} to 2.5×10^{-6} per °F for sawed members of different species and about 3×10^{-6} per °F for plywood. The average value is thus about one-sixth the value for aluminum and less than one-third the value for steel. Moisture is the greatest threat to a wood deck's dimensional stability. Under the extreme moisture variation anticipated in service, the expansion of plywood would roughly equal the expansion of steel under a 150°F temperature rise.

Moisture Absorption

Sooner or later, as an expanding gas, expanding freezing liquid, or contracting solid, moisture retained in materials like wood, concrete, and gypsum will damage the roof.

Opinions vary on the time required for the drying of poured structural decks. A minimum of 24 hr, depending on the weather, should elapse between the pouring of a gypsum deck and the application of roofing felts, according to some felt manufacturers and roofing contractors, who are concerned about the hazards of entrapping water-vapor released from inadequately dried deck surfaces. On the other hand, manufacturers of gypsum and bituminous roofing permit almost immediate coverage.

When to apply insulation or roofing to a poured gypsum or concrete deck can be simply resolved, under normal conditions, by the U.S. Army Corps of Engineers' test for dry deck. Pour a small amount of hot bitumen on the deck. If, after cooling, the bitumen can be readily removed with fingernails and hands, reject the deck as too wet for any application. If the cooled bitumen sticks to the deck, too tight to be removed by fingers, then accept the deck as dry enough for insulation or roofing application.

An alternative, or supplementary, test for properly vented poured concrete or gypsum decks is to apply hot bitumen as described above and observe for frothing or bubbling. If none occurs, the deck is probably dry enough for application.

Because of its high absorptivity, poured gypsum is not recommended for roof decks in buildings of high temperature and high relative humidity, e.g., laundries, bakeries, textile mills, etc. In highly humidified interiors wood and concrete are the most generally satisfactory roof deck materials. Properly maintained, to control corrosion, steel decks are also satisfactory.

Anchoring the Roof

The basic choice of techniques for anchoring built-up roofing or insulation to the structural deck is between nailing (or other mechanical fastening technique) and hot bituminous mopping or cold-applied adhesive. Roofing nails have large heads (about 1 in. in diameter) or caps through which they are nailed (see Figs. 3-4 and 3-5). The large heads are required to increase the perimeter of the felt tearing area, thus raising the tearing resistance of the felt (Chap. 9, "Wind Uplift").

For suitably dry deck surfaces, solid bituminous mopping is generally the best method for anchoring insulation, vapor barrier, or built-up membrane to the deck. Despite some exceptions to the rule, solid mopping generally provides the highest shear strength and resistance to wind uplift.

Secondary considerations, however, often rule out solid mopping as an anchoring method. On deck surfaces subject to shrinkage cracking, solid mopping increases the hazard of membrane splitting. By bonding two components throughout their contact area, solid mopping intensifies local stress concentration in a membrane directly over the deck crack. (Even when insulation board is interposed between the deck and membrane, solid mopping heightens the risk of splitting.) Solid mopping also seals in entrapped moisture, thereby promoting blisters and membrane wrinkling.

Strip or spot mopping alleviates both problems. By leaving areas unbonded to the membrane to distribute cracking strains, strip or spot mopping reduces membrane stress concentrations and possible splitting. Intermittent mopping provides lateral avenues of escape for water-vapor-pressure relief, which reduces the hazard of membrane blistering or wrinkling.

For poured gypsum decks, which are subject both to shrinkage and high moisture absorption, strip or spot mopping is preferred over solid mopping. But nailing is better yet, allowing both better stress distribu-

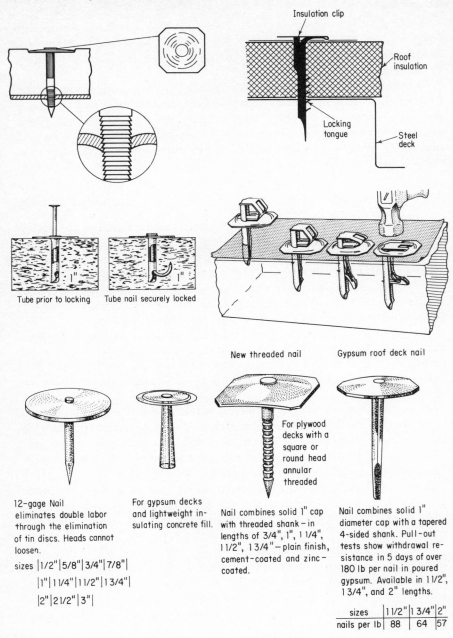

Insulation clip

Roof insulation

Locking tongue

Steel deck

Tube prior to locking   Tube nail securely locked

New threaded nail   Gypsum roof deck nail

For plywood decks with a square or round head annular threaded

12-gage Nail eliminates double labor through the elimination of tin discs. Heads cannot loosen.

sizes |1/2"|5/8"|3/4"|7/8"|
|1"|11/4"|11/2"|13/4"|
|2"|21/2"|3"|

For gypsum decks and lightweight insulating concrete fill.

Nail combines solid 1" cap with threaded shank – in lengths of 3/4", 1", 11/4", 11/2", 13/4" – plain finish, cement-coated and zinc-coated.

Nail combines solid 1" diameter cap with a tapered 4-sided shank. Pull-out tests show withdrawal resistance in 5 days of over 180 lb per nail in poured gypsum. Available in 11/2", 13/4", and 2" lengths.

sizes	11/2"	13/4"	2"
nails per lb	88	64	57

FIG. 3-4 *Roofing nails and mechanical anchors come in varied shapes and sizes. Small-headed nails are driven through minimum 1-in.-diameter steel disk caps, minimum 30-gage thickness.*

tion through the membrane and better access for water vapor pressure relief to vents.

Deck Surface and Joints

As a working surface for the application of insulation or built-up roofing, poured-in-place decks have an advantage that at least partly offsets the moisture problems; they provide large deck areas without joints.

FIG. 3-5 *Mechanical anchor is hammered through insulation and steel deck, which it grips with serrated surface. (ES Products, Inc.)*

Horizontal gaps between adjacent precast (or precut) units allow bitumen to drip through, creating a potential fire hazard. Vertical misalignment between adjacent units can ruin the surface as a substrate for applying the insulation, as can random cracking of poured decks (see Fig. 3-6).

There are several techniques for closing the horizontal gaps between prefabricated units and for leveling the ridges formed at joints. For wood sheathing, use of tongue-and-grooved instead of square-edged boards helps in both respects: the tongue-and-grooved joint is much

FIG. 3-6 *The large, structural crack in this poured gypsum deck must be investigated and corrected before installation of other roofing components can proceed.*

tighter than a square-edged butt joint, and the tongue-and-grooved boards do not warp as easily as square-edged boards. For plywood decks, when there is no supporting purlin or other structural member under the joint, use of H-shaped metal clips between adjacent units can prevent unequal deflection and thus avoid formation of vertical irregularities.

As precautions against bitumen drippage through the joints of wood decks, plywood may be stripped with felt, and board sheathing may be covered with 5-lb rosin-sized paper or 16-lb saturated felt. For glass-fiber felt membranes on nailed wood decks, one manufacturer recommends a glass-fiber "combination" sheet—an asphalt-coated base sheet with kraft paper laminated to the underside.

Joints in preformed mineralized wood fiber or precast concrete decks should be pointed with cement mortar, which is itself subject to shrinkage cracking (see Fig. 3-7). Prestressed concrete units bent in a convex deflection curve by the prestressing may require a concrete topping to level a surface marred by extreme joint irregularities. In less serious cases it may require only local grouting.

Low spots on concrete deck surfaces (more than $\frac{1}{2}$ in. below level) should be filled with portland cement mortar. High spots should be ground down.

Before insulation, built-up roofing, or vapor barrier are applied, poured concrete decks should receive a surface primer of cutback bitumen. The primer has a dual purpose: (1) to provide a surface film for adherence of bituminous mopping or cold adhesive; and (2) to absorb the dust that inevitably remains even after the surface is cleaned.

Lightgage steel decks are subject to longitudinal and transverse deflection (dishing). Deflection prior to application of cold adhesives or bitumen to steel decks can prevent good bond with the insulation placed on them. Live-load deflection *after* the insulation is placed can break a bond originally attained if the deck and insulation do not deflect in perfect congruity. As a safeguard against the loss of bond at the deck-insulation interface, the designer should limit vertical deviation from a plane deck surface between deck ribs to $\frac{1}{16}$ in. And as a safeguard against local distortion and breaking of bond, he should specify minimum 22-gage steel deck (see Chap. 9, "Wind Uplift").

ALERTS

Design

1. Check deck and its supporting structure for ponding deflection. Include factor for creep or plastic flow in computing deflection of concrete or wood members.

FIG. 3-7 *Shrinkage cracking of grout over bulb tees supporting preformed structural wood-fiber deck (top) requires special attention for anchorage of components to avoid membrane splitting (discussed in Chap. 6, "Elements of the Built-up Roofing Membrane"). When built-up membrane is applied direct to deck, base sheets should be nailed (bottom).   (GAF Corporation.)*

2. Design the roof to drain, with $\frac{1}{4}$ in. per ft minimum slope.

3. Check method of attachment of next component above deck with insulation manufacturer, roofing manufacturer, and insurance agent (see Chaps. 8 and 9, "Fire Resistance" and "Wind Uplift").

4. To reduce the threat of bitumen drippage when insulation, vapor barrier, or base sheet is hot-mopped to the deck, specify ASTM Type III or Type IV asphalt.   Specify rosin-sized paper, felt, or tape over joints in wood sheathing, plywood, or other prefabricated units to be nailed.

5. Check fire resistance of entire roof-deck-ceiling assembly (see Chap. 8, "Fire Resistance").

6. Limit steel decks to minimum thickness of 22 gage.

7. Recommend minimum $\frac{7}{16}$-in.-diameter roofing nail heads, to be driven through metal caps of minimum 1-in. diameter and minimum 30-gage thickness.

Field

1. Require a smooth, plane deck surface, with proper slope. For prefabricated deck units, tolerances are:

> Vertical joints:—$\frac{1}{8}$-in. gap
> Horizontal joints:—$\frac{1}{4}$-in. gap
> Flat surfaces of steel decks:—$\frac{1}{16}$ in.
> (between adjacent ribs)

FIG. 3-8 *Decks should be checked for equipment wheel loads, which may deflect steel decks, in particular, and break the bond between deck and insulation.*

When the vertical joint tolerance is exceeded, require leveling of the deck surface on a maximum 1-in.-per-ft slope with grout or other approved fill materials.

When the built-up membrane is applied direct to the deck, require caulking or stripping of all joints. Where insulation is applied on monolithic poured-in-place decks, fill the low spots (more than $\frac{1}{2}$ in. below true grade) and grind high spots. Openings over $\frac{1}{4}$ in. in wood decks should be covered with nailed sheet metal.

2. Permit no deformed side laps, broken or omitted side welds in metal decks.

3. Limit moisture content in deck materials to satisfactory levels—for example, 19 percent for wood and plywood.

4. Require waterproofed tarpaulin coverings over moisture-absorptive deck materials (wood, preformed wood fiber, precast concrete, precast gypsum) stockpiled at the site. Require also that these materials be blocked above the ground.

5. Prohibit placing of vapor barrier, base sheet, insulation, or roofing on deck containing water, snow, or ice. Decks must be clean and dry (including flute openings of metal decks) before next roofing component is applied. A test, useful under normal circumstances, for dry gypsum or concrete deck surface to be hot-mopped is as follows: Pour a small amount of hot bitumen on deck. If the bitumen, after cooling, can be removed with fingernails and hands, reject this deck for application of insulation. If the bitumen sticks to the deck, and can't be removed by the fingers, then the deck is dry enough to receive the mopping. An alternative or supplementary test for poured-in-place concrete or gypsum decks is to apply hot bitumen, as above, and observe for frothing or bubbling. If none occurs, the deck is dry enough.

6. Place plywood walkways on a vulnerable deck surface—e.g., light-gage steel.

7. Forbid stacking of materials on deck in piles that exceed design live load. Check also for equipment wheel loads.

8. Set close tolerances for steel-column base plate elevations (because they determine roof framing and deck slope).

FOUR

Vapor Control

Vapor barriers are essentially impermeable membranes designed to retard the flow of water vapor into the insulation and built-up membrane. Condensation of this vapor can impair the thermal resistance of the insulation and ultimately destroy the insulation itself. Liquid moisture may collect at the underside of the insulation until it flows into cracks or joints in the deck and leaks onto a suspended ceiling or directly into a room. If trapped condensate reevaporates, it can build up pressures approaching 400 psf. The resulting blisters can delaminate or tear the built-up roofing plies, destroying the weatherproofing cover provided by the membrane. The trapped condensate can also freeze and expand, breaking the bond between insulation and base coat, or even breaking through the built-up roofing membrane itself.

Paradoxically, more efficient insulation heightens the problems of condensation. Modern insulating materials separate such widely varying atmospheric conditions between the interior and exterior that they promote condensation within the built-up roofing system. Insulation shifts the dew point (the surface temperature at which water vapor will condense) from *under* the roof system to *within* the roof system (see Fig. 4-1). Thus, other factors being equal, the more efficient the insulation, the more need for a vapor barrier, subroof ventilation, or other means of preventing migrating water vapor from condensing in the built-up roof.

VAPOR MIGRATION

In the generally temperate climate of the United States, the normal direction of vapor migration is upward through the roof from a heated interior toward a colder exterior (from high to low vapor pressure). For refrigerated cold-storage buildings, however, or for airconditioned buildings in summer, this direction is reversed, with vapor migration proceeding downward from the warmer and normally more humid exterior toward the colder, drier interior.

The most common and most severe vapor migration problems occur during Northern winters, when temperature differences of 80°F or more between inside and outside can produce vapor pressure differentials of 30 psf or so. During the hottest summer weather, the extreme temperature difference between indoor and outdoor air drops to about 25°F (100°F outside and 75°F inside). Accompanying this lower summer temperature difference is a lower vapor pressure differential and consequently reduced rate of vapor migration. Moreover, since a built-up roof membrane in good condition is itself a vapor barrier, the summer condition normally requires no special attention.

Water vapor penetrates a built-up roof system through air leakage and diffusion. Depending on local conditions, air leakage and diffusion will vary in relative importance, but they will normally reinforce one another; i.e., both will tend to force the vapor in the same direction. In the normal case—heated, humid interior and cold, dry exterior—the

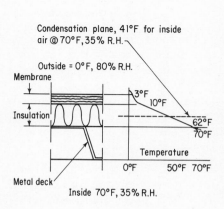

Condensation plane, 41°F for inside air @ 70°F, 35% R.H.

Outside = 0°F, 80% R.H.

Membrane

3°F
10°F

Insulation

62°F
70°F

Temperature

0°F   50°F 70°F

Metal deck

Inside 70°F, 35% R.H.

FIG. 4-1 *When insulation is used, the dew point usually shifts from* below *the roofing system to* within *the roofing system. The dew point of the inside (70°F, 35 percent relative humidity) is 41°F. Thus migrating water vapor will condense within the permeable insulation somewhere between the 41°F plane and the insulation-membrane interface. Without insulation, water vapor would condense on the steel-deck surface at about 35°F.*

atmospheric pressure under the roof will exceed the atmospheric pressure on the roof, and the air escaping through deck joints and other small openings will convey water vapor into the roof system from below. Because warm air can hold more water vapor than cold outside air, the water-vapor pressure of a heated, humidified interior will exceed the outside vapor pressure and thus promote outward, upward vapor migration.

Yet the two mechanisms are physically independent. It is possible, for example, to have an inward air leakage associated with outward vapor diffusion. On a hot, windless day in an arid climate, a cooled, highly humidified plant might produce this phenomenon. Because of the reverse "chimney effect" created by the cooled interior, the outside atmospheric pressure could exceed the interior pressure, while the water vapor pressure imbalance from the humidified interior could produce upward water vapor diffusion. (For a more detailed discussion, plus vapor-migration calculation, see the section on "Theory of Vapor Migration" at end of chapter.)

SAFEGUARDS

The designer has two basic methods of preventing moisture penetration into the roof: (1) *general ventilation*, designed to carry interior moisture out on drafts of air circulating through a space left under the roof, and (2) *vapor barriers*, designed as highly impermeable membranes that prevent the infiltration of water vapor into the built-up roof system.

Disagreement on vapor barriers focuses on this question: Is a vapor barrier a practicable means of preventing water vapor entry, or should the designer accept water vapor entry as inevitable and design for its dissipation?

General Ventilation

Properly designed subroof ventilation is the most certain technique for preventing water-vapor infiltration into a built-up roof. But for low-sloped roofs, it poses the most difficult set of design conditions to meet. Because natural convection decreases with diminishing roof height, ventilation is far less effective for low-sloped roofs than for steeply sloped roofs. Thus under a level roof the mechanisms left for dissipating moisture are diffusion and wind-induced ventilation. Exterior wind-produced pressure differentials tend either to force moist air out of the loft space or to displace it with colder, drier air, but with minimal benefits, particularly for a large level roof area.

There are, moreover, practical obstacles to general ventilation in commercial, industrial, and residential apartment construction. In these buildings, ducts and pipes located in loft space pierce the ceiling, thus making an effective air seal difficult. Since warm, humid air will probably infiltrate the cold loft space and condense on cold surfaces, ducts and pipes must be insulated. Especially in multistory buildings, where the "chimney effect" increases pressures in upper parts of the building and promotes upward air leakage, the ceiling air seal becomes essential. A further practical objection to general ventilation is the increased building height required to provide the ventilation space. General ventilation is thus the highest first-cost method of controlling vapor migration. As a compensating advantage, however, it is the most foolproof method of controlling vapor migration.

Vapor Barrier

According to conventional vapor barrier theory, the vapor barrier forms an essentially impermeable surface on the warm, humid side of the roof sandwich, blocking the entry of water vapor. Vapor barriers are used only in roofing systems where the insulation is sandwiched between the structural deck and the built-up membrane. A properly designed and installed vapor barrier can prevent condensation from forming within the built-up roofing system (see Fig. 4-2).

Water-vapor Transfer   To qualify as a vapor barrier, a material should have a vapor permeance rating not exceeding 0.2 perm. A material

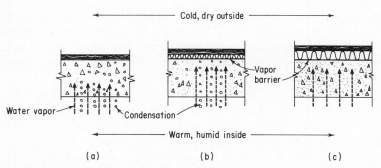

FIG. 4-2 *How a vapor barrier prevents condensation. In (a), an uninsulated concrete roof system, migrating water will condense when it reaches the dew point somewhere in the roof cross section. In (b), an* inadequately *insulated system in which the vapor barrier is at or below the dew point, migrating water vapor again condenses and drips. In (c), an* adequately *insulated system, the vapor barrier temperature is above the dew point and there is no condensation.*

rated at 1 perm admits 1 grain of water vapor per hour through 1 sq ft of material under a pressure differential of 1 in. of mercury (0.491 psi). Vapor resistance is the reciprocal of permeance.

For perm ratings of some vapor-barrier materials, refer to the ASHRAE (American Society of Heating, Refrigerating, and Airconditioning Engineers) "Guide and Data Book." Due to variations in some composite materials and other difficulties in measuring vapor-transmission rates, perm ratings are generally imprecise. Thus the sophisticated designer will not rely on overly refined calculations, but will design vapor barriers conservatively.

Another index of a vapor barrier's permeability, the *perm-inch*, or permeance of unit thickness, sometimes enables the designer to make a direct comparison between materials of equal thickness. (For films of less than 0.020 in. thick, the perm-inch index is impracticable.)

For many uses the 1-perm rating formerly set as the maximum permeance permissible for a vapor barrier is not good enough. Under a high differential vapor pressure, or even under low differential pressure over long time intervals (as in a refrigerated warehouse), a vapor barrier rated at 0.2 perm would admit too much water vapor. For certain kinds of industrial buildings subjected to continual high temperatures and high relative humidity, vapor barriers of less than 0.1-perm rating are recommended. In severely cold Northern locations, where the vapor pressure differential may persist in the same upward direction for weeks, a virtually impermeable vapor barrier is needed to prevent a destructive moisture build-up within the built-up roofing system.

Vapor-barrier Materials   An 0.2 perm rating is attainable by the most common vapor barrier—two or three moppings of bitumen and two plies of felt. A modern version of this vapor barrier uses one coated base sheet in combination with one or two bituminous moppings. Other materials that may qualify as vapor barriers include some plastic sheets (vinyl, polyethylene film, polyvinyl chloride sheets), black vulcanized rubber, glass, aluminum foil, and laminated kraft paper sheets with bitumen sandwich filler or bitumen-coated kraft paper. Steel decks with caulked joints may also qualify as vapor barriers (see Table 4-1).

As another layer in the multideck roofing sandwich, the vapor barrier must be solidly anchored to the deck below and to the insulation above. It must resist wind-uplift stresses and horizontal shearing stresses produced by thermal stresses in the membrane and transferred through the insulation, or produced by thermal or moisture-induced swelling or shrinkage in the insulation itself.

Vapor-barrier application varies with the materials. Sheets may be mechanically attached to the deck with nails or other deck-puncturing

table 4-1   PERM RATINGS OF SOME
VAPOR–BARRIER
MATERIALS*

(*Based on table provided by The American
Society of Heating, Refrigerating, and Air Conditioning Engineers, Inc.*)

	Permeance	
Material	Dry cup	Wet cup
Aluminum foil, 1 mil..............	0†	
Aluminum foil, 0.35 mil...........	0.05†	
Polyethylene, 4 mil...............	0.08†	
Polyethylene, 6 mil...............	0.06†	
Polyester, 1 mil..................	0.07†	
Saturated and coated roll roofing.....	0.05‡	0.24‡
Reinforced kraft and asphalt-laminated paper...........	0.3‡	1.8‡
Asphalt-saturated and coated vapor-barrier paper.........	0.2–0.3†	0.6‡
15-lb tarred felt..................	4.0‡	18.2‡
15-lb asphalt felt.................	1.0‡	5.6‡
Asphalt (12.5 lb/sq)...............	0.5‡	
Asphalt (22 lb/sq)................	0.1‡	
Built-up membrane (hot-mopped)....	0 ‡	

* Values are from ASHRAE "Handbook of Fundamentals," 1967 ed., chap. 19, table 1.

† Per ASTM E96-66 ("Water Vapor Transmission of Materials in Sheet Form").

‡ Per ASTM C355-64 ("Water Vapor Transmission of Thick Material").

devices, bonded with cold adhesives, or mopped with steep asphalt. The bonding agent used to anchor the vapor barrier to the deck is frequently used to anchor the insulation to the vapor barrier (see Fig. 4-3).

The vapor-barrier material must satisfy secondary requirements, such as practicable installation. Though they have low permeance ratings, plastic sheets, for example, create field problems that reduce their overall performance as components in the roof system. On windy days, these light, flexible sheets are difficult to install. Billowing and fluttering heighten the risk of tearing and make it difficult to flatten the sheet on the deck. Once in place, these sheets are vulnerable to traffic damage.

In addition to the foregoing, some plastic sheets have other disadvantages. Their shrinkage (after application) may tear the material where it spans across steel deck flutes. Hot bitumen leaking through the

FIG. 4-3   *This mechanical applicator spreads ribbons of cold adhe-sive on the steel deck, lays the plastic vapor-barrier sheet, and then spreads another set of ribbons on top to bond the insulation.   The adhesive ribbons should run parallel to the deck flutes.   The workman allows a time lag between the laying of the adhesive and the placing of insula-tion, to permit the volatile solvent to escape and insure better adhesion.   (Lexsuco, Inc.)*

insulation joints when the roof membrane is being applied may soften the sheets.   Bonding thin sheets to the insulation can pose difficulties.   The cold adhesive is normally applied parallel to the ribs of a steel deck on the plane surface.   Thus the rigid insulation, which spans from high point to high point of the rib, may make inadequate contact with the adhesive in a deck depression, resulting in poor attachment to the deck. This flaw can reduce the system's wind-uplift resistance.   It can even reduce the membrane's splitting resistance, which depends on the transfer or horizontal shearing stresses through the membrane-insulation interface and the insulation-deck (or vapor-barrier) interface.

Vapor-barrier Controversy   In the past, vapor barriers have been recommended in (1) areas where the January temperature averages 40°F or less, or (2) wherever building occupancy or use creates high relative humidity.

This approach is oversimplified.   Each project requires its own analysis.   Based on much sad experience, the efficacy of vapor barriers has come under attack from researchers and engineers administering vast building construction programs.   A new consensus developing among many roofing experts reverses the conventional policy on vapor barriers. According to conventional vapor barrier theory, the advice is, *"If in doubt, use a vapor barrier."* But according to the more sophisticated modern view, the advice is, *"If in doubt, omit the vapor barrier."*

Behind this skeptical view on vapor barriers lies a wealth of field experience, followed by successful changes in practice and more rigorous analysis of moisture problems. Plagued with numerous blistered roofs in its vast building program, one multiple building owner revised its former policy of generally using vapor barriers. This company now generally omits vapor barriers in buildings located south of 41°N latitude and where interior relative humidity is less than 45 percent. As a consequence, blistering has drastically declined; in fact, of 100,000 squares of built-up roofing installed within the past decade under the new policy, there have been no reports of blistering, which formerly constituted one-fourth of the company's roofing failures.

This successful policy reversal springs from a reevaluation of what is practicable in the field, not from a challenge to the basic theory of vapor migration. The key question is whether vapor barriers can be built soundly enough to form the virtually impermeable membrane required to do the assigned job. Field threats to the vapor barrier's integrity—punctures and cutting of new roof openings—make the construction of highly impermeable vapor barriers extremely difficult.

An inefficient vapor barrier is not merely inefficient; it becomes a definite liability. The ideal vapor barrier would be a jointless material with one-way permeability—allowing water and water vapor to leave but not enter the roof sandwich. Lacking such a miraculous material, the roof designer must accept the vapor barrier's liabilities along with its assets. Together with the built-up membrane, the vapor barrier seals entrapped moisture in the insulation instead of allowing it to escape. This trapped moisture, which may eventually migrate upward, can wrinkle the felts. Under high temperatures, it can expand and blister the built-up membrane; or it may cause extreme shrinkage and expansion in the insulation, resulting ultimately in membrane splitting. (For detailed analysis of moisture-caused membrane failure, see Chap. 6, "Elements of the Built-up Roofing Membrane.")

In summary, here is the case *for* the use of a vapor barrier:

▪ A vapor barrier can insure the continued thermal resistance of insulation sandwiched between the vapor barrier and the built-up membrane.

▪ A vapor barrier provides a good safeguard against vapor migration in case a building's use changes from a "dry" use to a "wet" use.

Here is the basic case *against* the use of a vapor barrier:

- The vapor barrier, together with the built-up roofing membrane, inevitably seals within the roof sandwich entrapped moisture that can eventually destroy the insulation, help split or wrinkle the built-up membrane, or, in gaseous form, blister it.
- In event of a roof leak through the membrane, the vapor barrier will trap the water below the insulation and release it through punctures that may be some lateral distance from the roof leak, thus making its discovery more difficult. A large area of insulation may be saturated before the punctured membrane can be repaired.
- A vapor barrier is a disadvantage in summer, when vapor migration is generally *downward* through the roof. (Hot, humid air can infiltrate the roofing sandwich through the vents, or through diffusion through the roof. It may condense on the vapor barrier itself.)
- A vapor barrier may be the weakest horizontal shear plane in the roofing sandwich. Failure at the vapor-barrier-insulation interface can result in splitting of the membrane. At the least, the vapor barrier introduces an additional component whose shear resistance may be critical to the membrane's integrity.

To permit use of a vapor barrier as a temporary roof is to invite roofing failure. Under the best conditions, installing an effective vapor barrier is a difficult job; but to subject it to the punishment of days or weeks of roof traffic and field operations makes a difficult job virtually impossible.

Venting the Insulation

Roof venting actually refers to the venting of one component—the insulation.

The vapor barrier protects only against future water-vapor infiltration. If the insulation or felts are wet when installed, or if insulating concrete is not allowed sufficient time to dry out before application of the built-up membrane, the vapor barrier works against the designer by sealing moisture within the roof. Though it probably will not dry out the insulation, venting is essential to relieve vapor pressure and reduce the hazard of blistering. Some experts claim that roofs with vapor barriers should always be vented, to provide escape paths and pressure relief against future moisture penetrating the roof.

There are two basic ways of venting the insulation:

1. *Stack venting* consists of vertical pipes opened to the outside air and shielded from rain by a conical cap plate, normally spaced at the grid intersection of 30-ft squares [see Chap. 5, "Thermal Insulation" (Venting)].

2. *Edge venting* creates horizontal escape paths through grooved edge boards open to the edge of the insulation and placed around the roof's

periphery [see Chap. 5, "Thermal Insulation" (Venting)]. Edge venting is, however, relatively ineffective for large roofs, with sizable roof areas more than 100 ft from the edges. For roofs over 40 ft wide, stack venting should supplement edge venting.

ALERTS

Design

1. As a basis for deciding whether or not to use a vapor barrier, calculate the location of the dew point and the rate of vapor migration under the worst winter condition (see example at end of chapter).

2. A vapor barrier on a roof destined for future penetrations is highly vulnerable to damage and likely to fail.

3. If in doubt, *don't* specify a vapor barrier. Use a vapor barrier only if a study of your conditions indicates a positive need for it. Normally, a vapor barrier should not be necessary unless the interior relative humidity exceeds 40 percent and January temperatures average less than 35°F. (For typical winter relative humidities under different occupancies, see Table 4-2.)

4. If unhumidified, exterior air is drawn into the building, a vapor barrier will normally not be required, unless the interior relative humidity is 60 percent or more.

5. In conjunction with a vapor barrier, specify an insulation that under prolonged exposure to moisture will (a) permit lateral transfer of vapor for venting pressure relief; and (b) retain sufficient strength despite the absorbed water. Specify a water-resistant method of bonding the insulation to the vapor barrier.

6. Never specify a vapor barrier between a poured structural deck and poured insulating concrete fill. (It will seal in moisture.)

7. Do not specify a vapor barrier in the roof over an unheated interior.

8. Consider an impervious, closed-cell board insulation in place of a vapor barrier.

9. Beware of specifying a vapor barrier over a wood deck. (Since it is generally impractical to furnish sufficient insulation to shift the dew-point plane from the wood into the insulation, the vapor barrier may be worse than useless.)

10. Check the vapor barrier's ability to take nailing or other anchorage punctures while maintaining a satisfactory perm rating.

11. Check a vapor barrier for its effect on the roof assembly's fire rating (see Chap. 8, "Fire Resistance").

12. For vapor barrier hot-mopped to deck, see Chap. 3, "Structural Deck" (Design Alert No. 4).

table 4-2   WINTER RELATIVE HUMIDITIES*

Dry occupancies:

Relative humidity expected to range under 20 percent, closely related to prevailing outdoor relative humidity and indoor temperature

Aircraft hangars and assembly plants (except paint shops)
Automobile display rooms, assembly shops (except paint shops)
Factories, millwork, furniture (except plywoods and finishing units)
Foundries
Garages, service and storage
Shops, machine, metalworking (except pickling and finishing)
Stores, dry goods, electrical supplies and hardware
Warehouses, drygoods, furniture, hardware, machinery and metals

Medium moisture occupancies:

Relative humidity expected to range 20 to 45 percent, varying with outdoor relative humidity but with moisture content increased by indoor activities, equipment, and operations

Auditoriums, gymnasiums, theaters
Bakeries, confectioners, lunchrooms
Churches, schools, hospitals
Dwellings, including houses, apartments, and hotels (highest relative humidity in kitchens, baths, laundries)
Factories, general maunfacturing, except wet processes
Offices and banks
Stores, general and department.

High moisture occupancies:

Relative humidity over 45 percent, determined primarily by processes, not by climate

Chemical, pharmaceutical plants
Breweries, bakeries, food processing, and food storage
Kitchens, laundries
Paint and finishing shops
Plating, pickling, finishing of metals
Public bath and shower rooms, club locker rooms, swimming pools
Textile mills, paper mills, synthetic fiber processing plants
Photographic printing, cigar manufacturing

---

* For more detailed recommendations, consult the ASHRAE "Handbook of Fundamentals."

Field

1. Do not use the vapor barrier as a temporary roof.
2. Where cellular foamed glass insulation is substituted for a vapor barrier, place board with tight joints.
3. Prohibit installation of light, flexible plastic vapor barriers on windy days.
4. Check steel decks for deflection, transverse dishing, or other irregularities that could prevent bonding of the vapor barrier to cold-applied adhesive.
5. Check spreader-applied ribbons of cold adhesive for proper consistency, thickness, and height.

THEORY OF VAPOR MIGRATION

Diffusion

Water vapor infiltrates a roof through two mechanisms: *diffusion* and *air leakage.* Compared with heat transmission, diffusion corresponds roughly with conduction; air leakage corresponds precisely with convection.

Diffusion can be explained by the kinetic theory of gases—a theory whose basic principles account for changes of state from liquids (or solids) to gases and the accompanying energy changes. In a sealed container of air with water in the bottom, at any temperature above absolute zero, some liquid (or solid) molecules will tear loose from the surface and escape into the air above. If constant pressure is maintained within the sealed container, then for each temperature a different quantity of water vapor (gaseous water molecules) will escape from the liquid and diffuse through the air.

So long as the water supply holds out, the liquid surface will reach equilibrium soon after the mixture reaches a stable temperature, with as many water molecules plunging back into the liquid as escape from its surface. When equilibrium is reached, the atmosphere is "saturated"; that is, the air-vapor mixture contains as much water vapor as it can retain at that temperature. The graph of Fig. 4-4 displays this capacity of air at constant pressure to hold varying quantities of water vapor at different temperatures. At normal atmospheric pressure of 14.7 psi and temperature of $-10°F$, a cubic foot of air can hold 0.3 grain of water vapor; at 70°F, it can hold 27 times that amount, or 8.1 grains.

So long as the liquid water holds out, the air-vapor mixture will remain saturated. If, however, all the water evaporates and the temperature rises, the water-vapor content of the air mixture will represent something less than the saturation limit. The ratio of the weight of water vapor

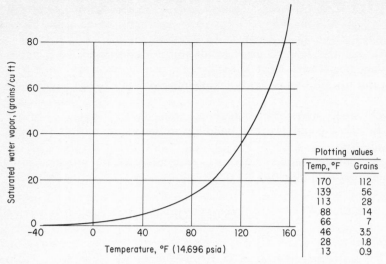

FIG. 4-4   *The water-vapor content in an air-vapor mixture rises rapidly with increasing temperature.*

Plotting values	
Temp., °F	Grains
170	112
139	56
113	28
88	14
66	7
46	3.5
28	1.8
13	0.9

actually diffused through the mixture to the water vapor in saturated condition is the *relative humidity*.   Thus, if liquid water is removed from a saturated air-vapor mixture at 46°F, with its 3.5 grains per cubic foot, and the temperature raised to 66°F, at which temperature the air can contain 7.1 grains per cubic foot, the relative humidity would drop from 100 to 49 percent.   Meanwhile, the *specific* or *absolute humidity* would remain constant at 3.5 grains of water vapor per cubic foot.

Now consider the opposite condition—falling temperature.   If the temperature drops from 66 to 46°F, the relative humidity rises from 49 to 100 percent.   Some of the water vapor will condense on surfaces at or below 46°F, since the heat energy in the air at that temperature is insufficient to maintain 3.5 grains of water vapor per cubic foot.   (The condensing water vapor loses its heat of evaporation.)   Thus 46°F is the *dew-point temperature*, or simply *dew point*, of air at 66°F and 49 percent relative humidity.   The dew point is the highest temperature of surfaces on which condensation will appear.

When the air-vapor mixture is saturated with vapor, the air temperature and the dew point are the same.   But whenever the relative humidity is less than 100 percent, the dew point is, of course, lower than the air temperature.   For air at 70°F, 10 percent relative humidity, the dew point is 13°F.

Another phenomenon associated with the diffusion of water vapor throughout air explains the penetration of water vapor into the built-up roofing.   The diffused water vapor is actually low-pressure steam, possessing its high latent heat energy.   Like any other diffused gas, water vapor mixed with air exerts a pressure independent of the pressure exerted by the air.   (Since the vapor pressure varies in direct proportion with the vapor weight, relative humidity is also defined as the ratio of the actual vapor pressure to the saturated vapor pressure at constant temperature.)   In accordance with Dalton's law, the partial vapor pressure is directly proportional to the volume occupied by the water vapor.

If water vapor is supplied to a space, the vapor pressure rises rapidly with increasing temperature.   Absolutely dry air (i.e., air with no suspended water vapor whatever) confined in a space of constant volume will gain 2.2 psi (from standard atmospheric pressure of 14.7 to 16.9 psi) under a temperature rise from 70 to 150°F.   (In accordance with Charles' law, the pressure increases in direct proportion with the rise in *absolute* temperature—from 530° Rankine to 610° Rankine.)   If to each cubic foot of dry air we add 76 grains of water vapor, the total pressure will increase 5.6 psi under the same 80°F temperature rise.   The water vapor exerts a partial pressure of 3.4 psi, or 490 psf, more than half again as much as the partial pressure increase in the heated dry air.   Thus warm, humid inside air inevitably exerts an unbalanced vapor pressure impelling vapor migration *from* the warm *toward* the cold side of the roof (see Fig. 4-5).

A curious consequence of the law of partial pressures concerns the water vapor's tendency to flow from a region of high vapor pressure toward a region of low vapor pressure, *regardless of the total atmospheric pressure.* Thus even if the outside air pressure exceeds the inside air pressure, the water vapor will tend to diffuse through the built-up roofing system from a warm, humid interior toward a cold, dry exterior.   The roof merely impedes the natural tendency of both the air and the water vapor to diffuse throughout the atmosphere in equal proportions.   In our vast atmospheric system, buildings are a mere local accident, and in seeking to achieve a perfectly proportioned mixture, the gaseous molecules in the atmosphere may move through walls and roofs in opposite directions—

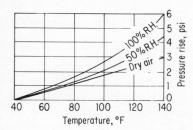

FIG. 4-5   *Pressure rise within a space of constant volume.*

with inward-bound air molecules passing outward-bound water-vapor molecules—as they seek their own partial-pressure level.

As an example of the pressure differentials encountered, consider a heated building interior within the normal range of heating and humidification—say, at 70°F and 40 percent relative humidity. Outside temperature is −10°F, 100 percent relative humidity (winter design conditions for Chicago). Under these conditions, 1 lb of indoor air contains 0.0063 lb (44 grains) of water vapor and 1 lb of outdoor air contains 0.0005 lb (about 3 grains) of water vapor. (These data are obtained from a psychrometric table or chart.) According to Dalton's law, the partial pressure of the water vapor is directly proportional to the percent by *volume* (not weight) of vapor. Since the air-water vapor mixture closely approximates an "ideal" gas, the vapor volume varies inversely with the ratio of the molecular weights of water vapor (18.02) and the air (28.96). Assuming the pressure of the air-vapor mixture, p_m = standard atmospheric pressure of 14.70 psi, we compute the vapor pressure p_v as a percentage of the air pressure p_a as follows:

$$p_v = \frac{28.96 \text{ (molecular weight of air)}}{18.02 \text{ (molecular weight of vapor)}} \times 0.0063 = 0.01 p_a$$

$$p_m = p_v + p_a = 14.70 \text{ psi}$$

$$p_m = 1.01 p_a$$

$$p_a = \frac{14.70}{1.01} = 14.5 \text{ psi}$$

$$p_v = 0.01 \times 14.5 = 0.145 \text{ psi} \times 144 = 21.1 \text{ psf}$$

We can thus assume for this example a vapor migration attributable to diffusion from the warm interior toward the cold exterior—a migration impelled under a pressure imbalance of 20.9 psf (obtained by subtracting the exterior vapor pressure of 0.2 psf from the 21.1-psf interior vapor pressure). As explained earlier, this vapor pressure imbalance induces an inside-out diffusion regardless of any possible atmospheric pressure imbalance forcing dry air toward the interior. In addition to the partial pressure difference, the rate of diffusion will depend on the length of the flow path and, of course, on the permeance of the media through which it passes (deck, vapor barrier, insulation, etc.).

Air Leakage

Recent research demonstrates that air leakage, which conveys water vapor with it, can be a greater factor than diffusion in accounting for vapor migration. If the components of the built-up roofing system—deck, vapor barrier, insulation, and built-up membrane—were perfect unpunctured surfaces, then diffusion would be the sole means of vapor migration.   But cracks, pinholes, and joint openings in deck, insulation, and flashing allow the passage of air from the high- to the low-pressure side of the roof.

Under the normally small pressure differences existing above and below the roof, the volume of air moved through these small openings is insignificant so far as it affects building heating and ventilation.   But a relatively small volume of air can transport a troublesome volume of water vapor.   At any point in the path where this migrating water vapor contacts a surface at or below the dew point, it will condense.

Several factors make the existence of atmospheric pressure differentials between inside and outside almost inevitable.   The major factors are temperature differences and wind.   Pressure differentials created by mechanical ventilation or exhaust systems are normally less important.

Chimney Effect   In a heated building the "chimney effect" produces higher pressures and consequent exfiltration in the upper part of the building, and lower pressures and consequent infiltration in the lower part of the building.   (Like a chimney, the heated building sets up a convection current, with cold air moving into the low-pressure space constantly vacated by the expanding, heated air, which rises and raises the pressure in the upper parts.)   In a cooled building the opposite occurs, with infiltration occurring at the upper levels and exfiltration at lower levels as the cooled, dense lower air presses out and sets up a downward convection current.

Wind Suction   Wind creates a positive static pressure on a windward wall in proportion to the wind's kinetic energy (velocity squared).   In conformance with the Bernoulli principle, the wind across a level or slightly pitched roof produces an uplift varying from a factor of $1.0p_v$ or more at the leading edge to $0.2p_v$ in some other region of the roof (p_v is the static pressure against a vertical surface).

As a hypothetical example of how pressure difference is created above and below the roof, consider a 400-ft-high building with a steady 20-mph wind blowing across the roof and a temperature differential of 60°F (70°F inside, 10°F outside).

At any level in the building, the pressure differential accompanying temperature difference can be computed from the following formula:

$$p_c = 7.6h \left(\frac{1}{T_o + 460} - \frac{1}{T + 460} \right)$$

$$p_c = 7.6 \times 200 \left(\frac{1}{70 + 460} - \frac{1}{10 + 460} \right) = -0.365 \text{ in. water}$$

where p_c = theoretical pressure (inches of water) attributable to chimney effect

h = distance (feet) from "neutral zone" (level of equal inside-outside pressure, estimated at mid-height for a multistory building without air-sealed floors)

T_o = outside temperature (°F)

T_i = inside temperature (°F)

Since static pressure for a 20-mph wind equals -0.193 in. of water, an estimate of $0.5p_v$ (for average roof suction) yields a static pressure of about -0.1 in. of water. Adding the two components of the pressure differential and multiplying by a conversion factor of 5.2 yields about 2.4 psf exfiltration pressure *upward* through the roof.

The practice of mechanically pressurizing airconditioned buildings, through a 10 to 20 percent excess of fresh supply air over exhaust air, can add another small increment of pressure imbalance to that produced by chimney effect and wind, further aggravating the problem of vapor migration. In arriving at his decision, the designer should weigh the benefits of mechanical pressurization against its liabilities. In humidified buildings he may even want to provide a small suction to reduce air leakage instead of deliberately increasing it.

Sample Vapor-barrier Calculations

Winter Condition   To illustrate a vapor migration problem for the normal (winter) condition, we shall assume a roof construction as shown in the cross section and the following data: outside temperature $T_o =$ 0°F, 90 percent relative humidity, and inside temperature $T_i = 70$°F, 40 percent relative humidity (see Fig. 4-6).

Since air-leakage computations can be made only for known, measured cracks and other openings—a clearly impossible task—the following computations will consider only diffusion. Because it is naive to assume great accuracy in such computations, pressures (in inches of mercury) will be carried only to the second decimal place. Due to many imponderables, vapor-migration computations are far less accurate than ordinary structural or mechanical design computations for, say, axial column stresses or heat losses. Thus vapor-migration computations are more an index to a problem than a precise design technique. In general, the designer must assume that at its worst, vapor infiltration will be greater than the computations indicate. As a mitigating factor, the worst condition is normally temporary.

The designer first tabulates thermal resistance (R values), perm values (M), and vapor resistances (perm value reciprocals) for the various roof components and air layers (see Fig. 4-6). The temperature loss upward

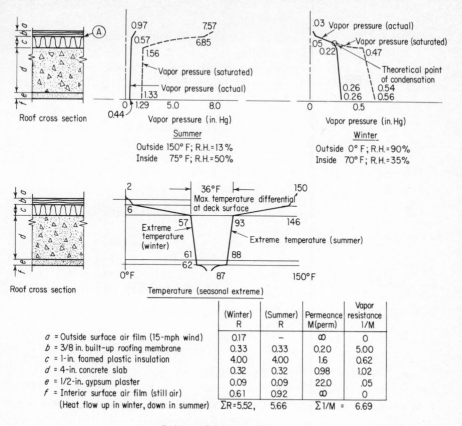

FIG. 4-6 *Vapor-barrier calculations, though imprecise, are sometimes necessary.*

through the roof cross section will be proportional to the sum of the R values of the components between the interior and a given plane, divided by the total sum of the R values.   Thus

$$T_x = T_i - \frac{\Sigma R_x}{\Sigma R} \left(T_i - T_o\right)$$

where T_x = temperature at plane X

ΣR_x = sum of thermal resistances from the interior to plane X

ΣR = total thermal resistances

The vapor-pressure gradient through the roof cross section can be computed in similar fashion. (It is analogous to drop in hydrostatic head in a series of canal locks.) Thus

$$p_x = p_i - \frac{\Sigma(1/M)_x}{\Sigma(1/M)}(p_i - p_o)$$

where p_x = vapor pressure at plane X

p_i = interior vapor pressure

p_o = exterior vapor pressure

$\Sigma(1/M_x)$ = sum of vapor resistances from interior to plane X

$\Sigma(1/M)$ = total vapor resistances

The critical plane is obviously A, the interface between the insulation and the built-up membrane (see Fig. 4-6). Because it lies on the far side of the relatively permeable insulation, it will represent the worst possible combination of low temperature and high humidity and thus pose the greatest risk for condensation.

$$T_a = 70 - \left(\frac{5.02}{5.52} \times 70\right) = 6°F$$

$$p_a = 0.26 - \left(\frac{1.69}{6.69} \times 0.23\right) = 0.20 \text{ in. Hg}$$

We can now calculate the vapor flow to and from plane A. When we check the saturated vapor pressure at A, however, we find that it is only 0.05 in. Hg, which is far below the calculated 0.20 in. Hg. Thus we know that condensation must take place at (or possibly below) plane A.

To calculate the vapor flow to plane A, we divide the pressure differential by the vapor resistance.

Vapor flow *to* plane A $\quad = \dfrac{0.26 - 0.05}{1.69} = 0.12 \text{ grain/(sq ft) (hr)}$

Vapor flow *from* plane $A = \dfrac{0.05 - 0.03}{5} = 0.004 \text{ grain/(sq ft) (hr)}$

These grains of water vapor are in transit through the roof, representing a more or less "steady state" of vapor migration, which will vary with changing temperature and humidity. Since the air-vapor mixture at plane A is saturated, all the migrating water vapor must condense. Thus we can compute the condensation rate merely as the difference between the vapor flow to plane A and the vapor flow from plane A— 0.12 grain/(sq ft)(hr).

To prevent condensation at plane A (the insulation-membrane inter-face), a vapor barrier must add sufficient vapor resistance on the high-pressure (warm side) of plane A to reduce the rate of vapor migration at plane A by at least 0.12 grain/(sq ft) (hr). Since the condensation rate (the excess of vapor flow *to* plane A over the vapor flow *from* plane A) must equal 0, we set the vapor flow to plane A equal to the vapor flow from plane A.

$$\text{Vapor flow to plane } A = \frac{0.26 - 0.5}{1.69 + X} = 0.004 \text{ grain/(sq ft) (hr)}$$

where X = required vapor resistance added by the vapor barrier.

$$0.21 = 0.004 \ (1.69 + X)$$
$$X = 53.5$$

$$\text{Required perm rating} = \frac{1}{X} = 0.02 \pm$$

To insure this virtually 0 perm rating is, to say the least, difficult. Thus the prudent designer will vent the roof in addition to providing a vapor barrier. In fact, if the ventilation could be designed to assure dissipation of the 0.12 grain/(sq. ft) (hr) (the computed condensation rate), the vapor barrier would not be necessary. But since the real world of construction differs so drastically from the ideal world of such calculations, a prudent designer confronted with such conditions would probably specify both a vapor barrier and some means of venting.

Summer Condition   In summer the problem of vapor migration is normally much less serious than in winter. Consider the same roof construction with outside temperature 100°F, 50 percent relative humid-ity; airconditioned interior at 75°F, 50 percent relative humidity (see Fig. 4-6). Assume further that the roof surface heats up to 150°F, reducing the relative humidity to 14 percent.

Computation of temperatures at the various planes between adjacent roof components indicates no planes at which condensation will occur, at least because of vapor migration resulting from diffusion. At plane A, the critical membrane-insulation junction under winter conditions, there is no problem if the built-up membrane is in sufficiently good condition to prevent extensive air leakage. The membrane's low insulating value combines with its low permeance (good enough to qualify the membrane as a good vapor barrier), to produce a high temperature (146°F) with a low relative humidity (8 percent) at plane A. Since the saturation vapor pressure remains well above the computed vapor pressure through the entire roof cross section—from the built-up membrane through the deck and the ceiling plaster—there is no problem of vapor-migration diffusion.

FIVE

Thermal Insulation

Roof insulation in an airconditioned or merely heated building offers the greatest return on initial investment of any building material. Yet, though insulation solves problems, it also creates problems; designing an insulated roof is far more complex than designing an uninsulated roof.

As part of its primary function, insulation creates greater interior comfort. In cold weather, it raises interior surface temperatures, thus preventing radiation losses that make room occupants uncomfortable. Roof insulation offers other benefits:

- It prevents condensation on interior surfaces.
- It may furnish a better substrate for application of the built-up roofing than the structural deck (e.g., steel).
- It stabilizes deck components by reducing their temperature variations and the consequent expansion and contraction.
- It can relieve stresses in the built-up membrane.

Accompanying these benefits are several liabilities:

- In reducing the probability of condensation on exposed interior surfaces, insulation increases the probability of condensation *within* a built-up roof system. (In heating season, insulation warms the interior surface and cools the exterior surface, thus increasing the chances that water vapor infiltrating from the warm side will condense within the roof cross section.)

Insulation raises roof surface temperatures in summer, thereby accelerating the oxidizing chemical reactions that harden asphalt surfacing and make it more susceptible to alligatoring, shrinkage; or other cracking.

• It produces more sudden thermal changes in the membrane (since the substrate conducts less heat to or from the roof surface), thus increasing the hazard of membrane splitting aggravated by thermal shock.

New insulating materials have raised many other problems—water absorption, drying rates, dimensional stability, treatment of joints, and choice of thickness. They make the selection of insulation one of the most complex tasks confronting the designer.

THERMAL INSULATING MATERIALS

The insulation used in built-up roof systems can be divided into four categories:

1. Rigid insulation, prefabricated into boards applied direct to the deck surface
2. Dual-purpose structural deck and insulating planks
3. Poured-in-place insulating concrete fills
4. Sprayed-in-place plastic foam

Soft blanket and batt insulations obviously cannot serve as the substrate for a built-up roofing membrane. They must be located below the structural deck, normally on the ceiling, below a ventilated space. This arrangement is especially good in arid climates with daily temperature extremes. The ventilated space interposes an additional insulating medium that modulates the ceiling temperature, thus increasing occupants' comfort. It can also serve to dissipate water vapor (see Chap. 4, "Vapor Control").

Below-deck insulation has several drawbacks, however, that normally outweigh its advantages. The additional building height required for ventilating space increases construction cost. And the below-deck arrangement sacrifices several other benefits of above-deck insulation: (1) it does not restrict thermal expansion and contraction in the deck, whose movement can damage the built-up membrane; (2) it does not provide an improved substrate for application of built-up membrane to a lightgage, ribbed metal deck; and (3) it does not protect utilities in the air space from freezing.

Rigid Board Insulation

Far more common than ceiling insulation in modern built-up roofing systems are the rigid board insulations, normally applied to the deck's top surface. These insulations are classified chemically as *organic* or *inorganic.*

Organic insulations include the various vegetable-fiber boards and foamed plastics. Inorganic insulation includes glass fibers, perlite board, cellular glass, mineral fiberboard, and poured insulating concretes with lightweight aggregates.

Physically, insulation is classified as cellular or fibrous.

Cellular insulation includes corkboard, foamed glass, synthetic rubber with gas-filled cells, and foamed plastics. The newest of these materials, the foamed plastics, are polystyrenes and, more recently, polyurethanes. Air, or other gas introduced into the material, expands it by as much as 40 times. Cells are formed in various patterns—open (interconnected) or closed (unconnected). Most rigid urethane foams are expanded with one of the halogen gases, which, due to their extremely low thermal conductivity, give foamed urethane its high insulating value. Gradual air leakage into the cells replaces the original gas and eventually reduces the thermal insulating quality by about one-third, but it nonetheless remains extremely good (see Table 5-1).

These materials come commercially in block sizes 12 × 18 in. to larger sheets 2 × 4, 3 × 4, or 4 × 4 ft in plan, in varying thicknesses.

Fibrous insulation materials include various fiberboards, which can be made of wood, cane, or vegetable fibers bonded with plastic binders. To make them moisture-resistant, they are sometimes impregnated, or later coated, with asphalt. Fibrous glass insulation consists of nonabsorbent fibers formed into boards with phenolic binders. It is surfaced with an asphalt-saturated, glass-fiber-reinforced organic material.

Dual-purpose Structural Deck and Insulating Plank

Made of cement-coated wood fibers set and cured in molds, preformed structural wood-fiber decks also serve as insulation. Physically, they are included among the fibrous organic insulations (see Chap. 3, "Structural Deck").

Poured-in-place Insulating Concrete Fill

The perlite or vermiculite aggregates normally used in lightweight insulating concretes contain varied-sized cells, which raise their thermal resistance to 14 to 20 times that of ordinary structural concrete.

table 5-1 ROOF INSULATION MATERIAL PROPERTIES[a]

Material	Organic (O) or inorganic (I)	Density, lb/cu ft	Thermal conductivity[b] k	Thermal resistance[b] $R = \frac{1}{k}$	Permeance[b], perm-in.	Coefficient of thermal expansion, in./in./°F $\times 10^{-6}$	Moisture expansion, % (50 to 97% relative humidity)	Compressive strength	Fire resistance
Cane- or wood-fiber board	O	15–22	0.36[c]	2.78[c]	20–50	3	0.5	Fair	Poor
Mineralized structural wood-fiber plank	O	22	0.60	1.67	Medium–High	...	...	Excellent	Fair–Good
Cellular glass	I	9	0.35–0.41[d]	2.86–2.44[d]	0	4.6	0	Excellent	Excellent
Preformed fiberglass board	I	4–9	0.21–0.26[d]	4.76–3.85[d]	Very high	...	<0.05	Poor	Good
Foamed polystyrene	O	1.9	0.22–0.26[e]	4.55–3.85[e]	Extruded 1.2 Bead type 2.0–5.8	35	...	Fair	Poor
Foamed urethane	O	1.5–2.5	0.17[e]	5.88[e]	1–2	30	...	Fair	Poor
Perlite aggregate board	I	11	0.38	2.63	25	Negligible	0.1	Excellent	Excellent
Lightweight concrete, perlite aggregate	I	20–40	0.70–1.15[f]	1.43–0.87[f]	High	4.3–6.1	...	80–450 psi	Excellent
Lightweight concrete, vermiculite aggregate	I	15–40	0.70–1.15[f]	1.43–0.87[f]	High	4.6–7.9	...	70–400 psi	Excellent
Corkboard	O	6.5–8.0, 12	0.25–0.28[d], 0.28–0.31[d]	4.0–3.57, 3.57–3.22	2.6	...	...	Fair	Poor

[a] Tabulated values are from varied sources. Designer should check with manufacturer and/or other sources for more precise values.
[b] See Glossary for definition.
[c] At 75°F.
[d] From 0 to 90°F.
[e] From 0 to 75°F.
[f] For 20 to 40 lb/cu ft range only (from ASHRAE "Handbook of Fundamentals," 1967 ed., chap. 26, table 3).

Perlite is silicaceous volcanic glass; vermiculite is expanded mica. Both materials expand 15 to 20 times at high temperatures somewhere between 1400 to 2000°F. Normally, the concrete made from these aggregates has a portland cement binder, but perlite aggregate sometimes has an asphalt binder (see Fig. 5-1).

Perlite aggregate is also used in gypsum decks to increase insulating efficiency.

Lightweight concrete fills are also produced in cellular form by adding a foaming agent to the mix.

Poured insulating concrete fills are an economical means of providing both insulation and a sloped roof surface for drainage. They also form a smooth substrate atop steel decks, precast concrete, and other decks for applying the membrane or supplementary insulation boards, if required. Unlike some other insulating materials, they offer excellent compressive strength and fire resistance.

As one notable advantage over lighter materials, massive insulating materials like lightweight concrete fill (which may weigh 50 times as much as foamed plastic of equivalent thermal resistance) stabilize the extreme fluctuations of insulated roof surface temperatures, thereby reducing the hazard of thermal shock (see Chap. 6, "Elements of the Built-up Roofing Membrane"). A roof membrane directly atop a light, highly efficient insulation reacts to the sudden heat of the sun like an empty pan on a burner. The metal in contact with a good insulator—in this case, air—heats up faster than the metal of a pan filled with a poor insulator like water. A massive insulating material, like the water in the heated pan, has greater heat capacity than a lighter, more efficient

FIG. 5-1 *Pumping lightweight insulating concrete mix to a steel deck (instead of conventional pouring) removes buggy and other equipment loads that could cause roof-damaging deflections. (Pittsburgh Corning Corp.)*

insulator.   Because they store and release heat more slowly than thinner, lighter insulations, these heavier materials tend to stabilize roof surface temperatures.   Thus the membrane temperature reacts less quickly and less intensively to the sudden heat of the sun emerging from behind a cloud or to the chill of a sudden rain shower.   Massive structural decks— notably concrete or poured gypsum—stabilize roof surface temperatures even more than insulating concrete.[1]

Sprayed-in-place Plastic Foam

A recent development in roof insulation is sprayed-in-place urethane foam.   This foamed-in-place plastic offers some advantages over the prefabricated boards—chiefly the elimination of insulation board joints that can form lines of stress concentration in the membrane.   Major drawbacks of the sprayed-in-place urethane are the difficulty in applying uniform thickness and achieving a level surface.

By combining two thermal-insulating materials, a designer can exploit their complementary qualities.   Foamed plastic insulation placed atop mineral fiberboard insulation or sloped insulating concrete fill (which provides roof drainage) creates a roof system with high insulating value and good fire resistance.   On a steel deck, such a system can qualify for a Class 1 fire-resistance rating, but not if the foamed plastic is placed directly on the steel deck (see Chap. 8, "Fire Resistance").

DESIGN FACTORS

The chief properties required of insulation are:

- Horizontal shear strength sufficient to maintain dimensional stability of the roof membrane under tensile contraction stress.
- Compressive strength sufficient to withstand traffic loads and hailstone impact.
- Cohesive strength sufficient to resist delamination under wind uplift.
- Resistance to damage from moisture absorption.
- Dimensional stability under thermal changes and moisture absorption.
- A surface absorptive enough to produce good bond strength with the bituminous mopping, but not so absorptive that it soaks up the bituminous coating.

Strength Requirements

To help maintain the dimensional stability of the built-up roofing membrane under contraction stresses, thereby preventing membrane

---

[1] For data on cellular and aggregate-type lightweight insulating concrete, see "Guide for Cast-in-place Low-density Concrete," Title No. 64-44, *ACI Journal*, September, 1967.

splitting, insulation requires good horizontal shearing strength. Without such help from the insulation, the membrane will almost inevitably split at points of stress concentration.

Overlooking this need for good anchorage at the deck-insulation interface, some designers of buildings planned for future expansion have specified insulation boards loosely laid on the deck, for easy removal when the roof is later converted to a floor. This first-cost economy may prove costly. An unanchored insulation substrate heightens the risk of membrane splitting or wrinkling. Through internal stresses produced by thermal and moisture changes, a built-up membrane exerts a rachet action on an unanchored, or even a poorly anchored, insulation. (The flexible membrane expands, compresses, and buckles in heat, and contracts and pulls in the cold, thus producing a cumulative rachet action toward the center of roof area.) Compounded by aging and moisture changes in the felts, this rachet action sometimes pulls the insulation 2 or 3 in. from the roof edges, destroying the edge flashing. Good anchorage between the deck and insulation is essential to resist these membrane-imposed stresses and to prevent splitting, especially in cold climates.

The orientation of insulation boards with respect to the felt rolls affects the membrane strength. The conventional practice of unrolling felts parallel to the continuous longitudinal joints of insulation boards maximizes the chances of splitting or wrinkling. The continuous longitudinal joints in the insulation are lines of maximum stress concentration, and the felts must resist these stresses in their weakest (transverse) direction (see Fig. 5-2).

Orienting the felt rolls perpendicular to the continuous longitudinal joint, as in Fig. 5-2*b* and *c*, matches maximum membrane stress with maximum felt strength, thereby minimizing the chances of membrane splitting or wrinkling.

Bond or adhesive strength, achieved through a bituminous mopping, cold-applied adhesive, nailing, or other mechanical fastening, is also important for resisting wind uplift. Bond strength is required at both the deck-insulation interface and the insulation-membrane interface.

Tensile strength to resist delamination under wind uplift is another important structural property of insulation. A minimum 30-psi compressive strength is recommended to resist traffic loads and hailstone impact.

The designer should require field tests to assure adherence to minimum specified cement content in lightweight insulating concrete. Traffic should not be permitted on a lightweight insulating concrete deck for at least 72 hr after pouring. (For methods of anchoring the built-up membrane to an insulating concrete substrate, see Chap. 6, "Elements of the Built-up Roofing Membrane.")

Moisture Absorption

Water entrapped in insulation (during roof application, through leaks in the membrane, or condensation of infiltrating vapor) can destroy the thermal-insulating value of some insulation materials.

Water vapor can flow wherever air can flow—between fibers, through interconnected open cells, or where a closed-cell structure breaks down. Water can easily penetrate the larger enclosures of an insulating concrete. Wherever water replaces air, insulating value drops drastically, since water's thermal conductivity exceeds that of air by 20 times.

Fibrous, organic insulations are especially vulnerable to moisture damage. Free water will eventually damage any fibrous organic material or organic plastic binder. Fiberboard long exposed to moisture may warp or buckle and eventually decay. The expansion and contraction that accompany the changing moisture content can seriously damage the built-up membrane.

Some preformed structural wood-fiber planks are subject to inelastic sag, attributable chiefly to moisture weakening of some (not all) cementi-

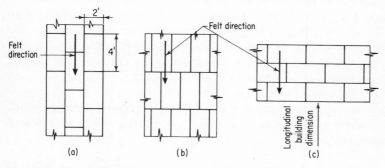

FIG. 5-2 *Insulation patterns. The conventional pattern of unrolling felts parallel to the continuous longitudinal joints of insulation boards (a) heightens the chances of membrane splitting or wrinkling. The continuous, longitudinal joints are lines of maximum stress concentration, and the felts must resist these stresses in their weakest (transverse) direction.*

In pattern (b), a compromise, the felts resist the maximum stresses produced by the insulation joints in their strongest (longitudinal or machine) direction. The longitudinal joints between insulation boards, however, become more critical and subject the felts to fairly high contraction stresses in their weak direction.

In pattern (c), the best, the longitudinal joint again works against the strong felt axis, and the contraction stresses of the continuous joints in (b) are reduced, further relieving the stresses in the felts' weak direction.

tious binders.   For that reason, these planks are not recommended for use over highly humidified interiors.

Though less vulnerable to moisture, inorganic materials are not immune.   Water penetrating into fiberglass insulation not only impairs the insulating value, but may also dissolve the binder.   Foamed cellular glass will neither absorb water in its closed cells nor allow the passage of water vapor.   But if moisture is trapped between the built-up roofing and the top of the insulation, water will accumulate in open-surface cells. With roof surface temperature alternating above and below freezing temperatures, ice formation can break down the walls between open-surface cells and the interior closed cells.   Repeated freeze-thaw cycles progressively destroy the foamed glass, leaving a water-saturated gray-black dust.   (Laboratory tests have indicated complete breakdown under as few as 20 such cycles.)

Foamed polystyrene cells may also break down under freeze-thaw cycles.   Compounding the problem is the threat to the built-up membrane, which can be destroyed by vapor-caused blisters, or to the ceiling, when condensed or leaking water drops out of the insulation.

Lightweight insulating concrete may contain high moisture content, even though it has been in place for years.   On some roofs in place for 10 or more years, laboratory tests of insulating concrete cores cut from the roof system have contained from 10 to more than 50 percent moisture by weight.   Though a roof leak may be suspected as the moisture source, the moisture may be condensate formed within the insulation, or may even be the result of long-term water retention.   Most lightweight insulating concrete fills are poured on vented lightgage steel decks, which thus admit water vapor from the building interior as well as expelling it into the interior.

Venting

Venting of insulation is always advisable when a vapor barrier is used. For poured concrete insulating fill poured on a monolithic structural concrete deck, venting is advisable even without a vapor barrier.   Vapor barriers offer no protection against moisture contained in insulation that was wet when installed; in fact, they help to seal it in.   And since vapor barriers and built-up roofing membranes are inevitably permeable to some degree, venting is recommended to relieve vapor pressure generated under a heated roof surface and to help dissipate moisture in poured insulation.

Insulation may be vented in three basic ways:

Stack Venting   Stack venting consists of vertical pipes opened to the outside air and shielded from rain by a cover (see Fig. 5-3).   Special vent ducts made of porous fiber materials are needed to conduct vapor through insulating concrete.

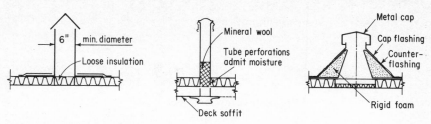

FIG. 5-3   *Typical stack vents.*

Edge Venting   Edge venting creates horizontal escape paths through grooved edge boards open to the edge of the insulation and placed around the roof's periphery (see Fig. 5-4).   On large roofs with interior areas more than 100 ft from the roof periphery, both stack and edge venting are advisable (see Chap. 4, "Vapor Control").

Underside Ventilation   Underside ventilation for lightweight insulating concrete on steel decks consists of continuous openings, $\frac{1}{8}$ to $\frac{3}{16}$ in.

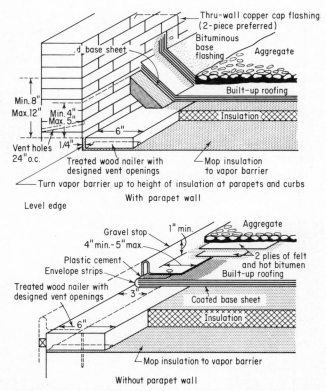

FIG. 5-4   *Typical edge-vent details.*   (*W. R. Grace Co.*)

wide, between adjacent deck units. Metal vent clips hold the joints open. Poured gypsum decks are vented through gypsum form boards.

Dimensional Stability

As the filler in the roofing sandwich between deck and built-up roofing membrane, insulation must retain stable dimensions or it may damage the built-up membrane. The extreme temperature ranges experienced by roofs—daily changes over 100°F and annual changes of 200°F—can turn some roof insulations into agents of destruction. Contraction and expansion with changing moisture content can also distort some insulation materials beyond tolerable limits. Repeated expansion and contraction of insulation boards may split the membrane.

Insulation made of cellulosic fibers swells with moisture absorption and contracts on drying as much as 0.2 to 0.5 percent, with changes in relative humidity from 50 to 90 percent. Under such a humidity change, a 4-ft-long cellulosic fiber board may contract ¼ in., thus opening a joint.

Insulation materials with a high coefficient of thermal expansion can contribute to the splitting of roof membranes subjected to extreme temperature changes (see Fig. 5-5). Splitting can occur when a roof membrane shrinks as temperature drops, or as wet felts dry out. Since the flexible membrane buckles under compressive stress, the membrane cannot expand with rising temperature, and so the contraction proceeds with a sort of ratchet action resulting in accumulating contraction. If its bond to the built-up roofing membrane is stronger than its bond to the deck (the normal condition), the insulation can be dragged across the roof by the contracting membrane, leaving a joint gap as wide as 3 in. between the wood deck nailer and the peripheral insulation. With its high deformation under thermal cycles, the insulation can open joints

FIG. 5-5 *Excessive contraction of preformed insulation helped to produce the membrane split shown above.*

elsewhere as it contracts, and this contraction can split the roof at interior locations (see Fig. 6-13). Because of its normally wider joints, thick insulation can aggravate this problem.

Under a 100°F daily temperature drop, a 4-ft-long polystyrene board will contract nearly $\frac{3}{16}$ in. Foamed urethane, with its thermal coefficient of 30×10^{-6} in. per °F, poses nearly as grave a problem as does foamed polystyrene.

Because of its somewhat unpredictable tendency to expand for long periods after its pouring, perlite insulating concrete fill requires expansion joints around the roof-deck perimeter and at skylights, penthouses, parapet walls, equipment base curbs, ventilators, and other roof-penetrating elements. These expansion joints should be at least 1 in. thick, and made of material that will compress to one-half its unstressed thickness under a stress of 25 psi.

The expansion joints are a safeguard against a highly complex chemical reaction that affects some (not all) perlite insulating concretes. Normally, this long-term expansion is offset by initial shrinkage.

Because its high initial shrinkage more than offsets any further expansion, vermiculite concrete does not normally require expansion joints.

Refrigerated Interiors

The architect should not attempt to economize by insulating a refrigerated interior with insulation placed atop the structural deck. There are two main reasons for avoiding this practice:

- Placing the roof insulation between a cold, dry interior and (in summer) a hot, humid exterior increases the pressure of water-vapor migration down through the membrane. If the refrigerated space is maintained at 0°F, this pressure may always be downward. Constant downward vapor migration will almost inevitably saturate the insulation and, at the least, threaten membrane blistering, icing delamination, and the other problems associated with entrapped moisture.
- The thicker insulation required for a refrigerated space creates a less stable substrate for the membrane than the normally thinner roof insulation. The inevitably wider joints and greater movement in a thickened substrate magnify the hazard of membrane splitting.

Providing a ventilated ceiling air space between the separately insulated refrigerated space and the roof deck relieves this threat to the roof. It greatly reduces the vapor-pressure differential between the roof and the interior. It also reduces the chances of water penetrating into the refrigerated space, where it will freeze, reduce refrigerating efficiency, and raise operating costs.

Application Problems

The foamed plastics have limitations that may offset their light weight and high thermal-insulating quality. Foamed polystyrene is combustible, and its cell structure starts to collapse at 175°F. Since hot bitumen is usually mopped to the surface at temperatures between 350 and 400°F, the roofing cannot be applied in the normal manner. One insulation manufacturer recommends hot-mopping of the deck and allowing the bitumen to cool before placing his insulation. Placing it in cooled bitumen may, however, weaken the bond. Once the insulation is installed, the application of the roofing is difficult. It can be accomplished, at increased labor cost, by mopping the underside of the first ply of roofing felt, and then placing it onto the insulation. Another technique calls for use of a coated base sheet laid dry over the insulation. When the top of the base sheet is hot-mopped, its bottom coating softens and produces the required bond at the top of the insulation. Still another technique features a drum-type felt layer that coats the back of the felt.

Foamed urethane has similar limitations, although its cell structure will not collapse until temperatures rise to about 250°F. The dimensional stability of the material is questionable, due to its reported growth with age. Short-term exposure to hot bitumen may cause no trouble at all, especially if the surface is protected by impregnated kraft paper. To protect foamed urethane from hot bitumen, at least two manufacturers have sandwiched a urethane core between asphalt-saturated and coated sheets, which also stiffen the insulation against dimensional changes.

Joint Taping

An unresolved industry disagreement concerns the taping of insulation joints (see Fig. 5-6). According to one insulation manufacturer, reinforced laminated paper tape applied over the joints of prefabricated plank insulation offers several benefits:

- It prevents the bitumen from running into joints.
- It increases resistance to water vapor access into the built-up roofing membrane.
- It provides additional resistance to damage to the built-up roofing, both during application and afterward.
- It increases the membrane's ultimate tensile strength.

Impartial experts confirm that joint taping can eliminate areas of stress concentrations, thereby strengthening the built-up membrane and preventing splitting.

FIG. 5-6 *Taped insulation joints—done manually on small jobs, mechanically on large jobs—can alleviate stress concentrations in the membrane.*

ALERTS

Design

1. If possible, specify insulation in one of five standard thermal conductances (C value = 0.12, 0.15, 0.19, 0.24, 0.36).

2. When practicable, align the continuous, longitudinal joints of insulation boards parallel to the *short* dimension of the roof and perpendicular to the rolled (longitudinal) direction of the felts.

3. Select minimum insulation thickness. Use two layers in preference to one. For multiple layers, specify staggered joints.

4. Do not overinsulate in climates characterized by severe winters or sudden temperature drops. (It promotes ice formation on the membrane and greater temperature extremes within the membrane.)

5. Investigate the need for taping insulation board joints (to relieve membrane stresses). Consider taped joints whenever dimensionally unstable insulation boards are specified. (Do not substitute insulation with taped joints for a ply of roofing felt.)

6. Require bituminous impregnation of wood fiberboard and other organic fiberboard insulations.

7. Install insulation for a refrigerated space below the roof deck.

8. Investigate the need for a vapor barrier to shield the insulation from water vapor migrating from a warm, humid interior (see Chap. 4, "Vapor Control").

9. Specify edge venting and/or vapor-relief vents in roof insulation over a vapor barrier, spaced uniformly, with one vent for each 900 sq ft.

10. For thermal insulation hot-mopped to the deck, see Chap. 3, "Structural Deck" (Design Alert No. 4).

Field

1. Require stockpiled insulation stored at job site to be covered with waterproofed tarpaulins, and raised above the ground.

2. Require a smooth, plane deck surface. [For specific tolerances, see Chap. 3, "Structural Deck" (Field Alert No. 1).]

3. Prohibit placing of insulation on wet decks or snow- or ice-covered decks. Decks must be cleaned and dried (including flute openings of metal decks) before insulation is installed. Normally, test for dry deck surface as follows. Pour a small amount of hot bitumen on deck. If the bitumen, after cooling, can be removed with fingernails, reject this deck for application of insulation; the moisture content is too high. If the cooled bitumen sticks to the deck, and can't be removed by the fingers, then the deck is dry enough to receive the mopping.

An alternate test for properly cured and dried cast-in-place decks (e.g., concrete, gypsum) is to apply hot bitumen, as above, and observe for frothing or bubbling. If none occurs, the deck is dry enough.

4. Limit moisture content to satisfactory levels, not to exceed the equilibrium moisture content.

5. Limit daily application of insulation to an area that can be covered on same day with built-up roofing membrane.

6. Place plywood walkways on insulation vulnerable to traffic damage.

7. Set adjacent units of prefabricated insulation with tightest possible joints. *Trim or discard units with broken corners or similar defects.*

8. Provide water cutoffs at the end of the day's work on all decks with insulation boards, cut along vertical face, and remove when work resumes (see Fig. 5-7).

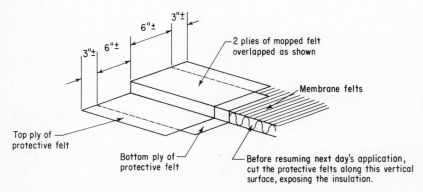

FIG. 5-7 *Temporary insulation-board cutoff detail (to be used at end of day's work).*

9. Use steep asphalt when hot-mopped bitumen is specified for bonding prefabricated insulation to the deck. Normally, insulation should be solid mopped.

PRINCIPLES OF THERMAL INSULATION

Heat is transferred through (1) conduction, (2) convection, and (3) radiation. *Conduction* depends on direct contact between vibrating molecules transmitting kinetic (or internal heat) energy through a material medium. *Convection* requires an air or liquid current, or some moving medium, to transfer heat physically from one place to another. *Radiation* transmits heat through electromagnetic waves emitted by all bodies, at an intensity dependent on their temperature.

Radiation accounts for the extremes in roof surface temperatures, above and below atmospheric temperature. Sun rays can raise the surface temperature of an insulated roof 75°F above the air temperature, and on a clear night, without cloud cover to reflect radiated heat back to earth, roof surface temperature can drop 10°F or more below the air temperature. Thus, in a climate with design temperature varying from a summer maximum of 95°F to a winter low of 0°F, the annual variation in roof temperature may be 160°F or more. Because it conducts less heat to or from the roof surface, insulation within the roof sandwich will *increase* the extremes in roof surface temperature; for a black roof surface the annual temperature differential could exceed 200°F. But insulation will greatly reduce the daily or annual temperature difference experienced by the deck.

Heat flows through building materials primarily through conduction, less so by convection. Through an air space all three modes of heat transfer are at work. Thermal insulation resists all three, but it primarily resists conduction.

Reflective insulation—normally with an aluminum foil or aluminum-pigmented heat-reflective coating—is common in walls with an enclosed air space. But the difficulty of accommodating such a space under a roof, plus the need to pierce the foil with numerous openings for pipes, conduit, ducts, and other mechanical items, makes reflective insulation generally impracticable for roofs. So does the difficulty of keeping reflective insulation from contacting other materials, thereby losing its effectiveness.

Good thermal insulators generally depend on the entrapment of air, a poor heat conductor, in millions of small cells or pockets; these act to arrest the transfer of heat by convection. Carbon steel, a good heat conductor, will transfer nearly 2,000 times as much heat as an equal thickness of air. Cellular insulations, e.g., foamed glass or plastic, acquire their thermal-insulating value by establishing a temperature gradient through the cross section, with each tiny cell of entrapped air making its contribution to the total resistance to heat flow.   Fibrous insulations exploit thin pockets of air between the fibers.

The efficiency of thermal insulation also depends on its capacity for impeding air flow, thus resisting heat flow by convection, and on the low thermal conductivity of its basic materials.

The quantity of heat transferred through a building component varies directly with the temperature differential, the exposed area, and the time during which the transfer takes place.   It varies inversely with thickness. A good thermal-insulating material is a poor thermal conductor (and vice versa); it also blocks the passage of air, thus resisting convective heat flow. If it is sufficiently opaque, it resists the penetration of heat radiation.

Heat Flow Calculation

The simplified heat transfer calculations suitable for building design require a knowledge of four indexes of heat transmission:

1. Thermal conductivity k = heat (Btu) transferred per hour through a 1-in.-thick, 1 sq ft area of homogeneous material per °F of temperature difference from surface to surface.   [The unit for k is (Btu)(in.)/(hr) (sq ft)(°F).]   To qualify as thermal insulation, a material must have a k value of 0.5 or less.

2. Conductance, $C = k/$thickness, is the corresponding unit for a material of a given thickness.   [The unit for C is (Btu)/(hr) (sq ft) (°F).]   For a 2-in.-thick plank of material whose $k = 0.20$, $C = 0.10$.

3. Thermal resistance, $R = 1/C$, measures a material's resistance to heat flow.   (A material with $C = 0.25$, denoting heat transmission of 0.25 Btu per hr through 1 sq ft of material per °F, has an R value of 4.0. It would take 4 hr to transmit 1 Btu through 1 sq ft per °F.)

4. Overall coefficient of transmission U is a unit like k and C, measured in Btu transmitted per hour through 1 sq ft of the construction per °F from air on one side to air on the other.   It relates, however, to the several component materials in a wall or roof. U can be calculated from the following formula:

$$U = \frac{1}{\Sigma R} \tag{5-1}$$

where R = sum of the thermal resistances of the components, plus the resistance of the inside and outside air films.

EXAMPLE: To calculate the insulation's required thermal resistance R_i, the designer usually starts with a target U factor set by the mechanical engineer.   For the other components, he merely tabulates the resistances

R for all the materials, including inside and outside air films. (For the summer condition, when roof temperatures often rise to 60°F or more above outside air temperature, it is prudent to assume a roof surface temperature of 150°F or so and to omit the outside air-film resistance.) Data for conductances of various materials are available from the ASHRAE "Handbook of Fundamentals." If not available in general tables, data for proprietary insulating materials should be furnished by manufacturers.

Consider the roof shown in cross section (Fig. 5-8) with a target U factor of 0.20.

By Eq. (5-1),

$$U = \frac{1}{2.60 + R_i} = 0.20$$

and

$$R_i = \frac{1}{0.20} - 2.60 = 5 - 2.60$$

requiring

$$R_i = 2.40$$

A 1-in.-thick cellular glass ($R = 2.56$) is satisfactory.

Omission of the insulation would increase the U factor as follows:

$$U = \frac{1}{5.13 - 2.56} = 0.39$$

Thus the 1-in.-thick cellular glass insulation halves the heat losses through this roof.

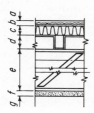

	Resistance R
a = Outside surface air film (15-mph wind)	0.17
b = 3/8 in. built-up roofing membrane	0.33
c = 1-in. cellular glass insulation	R_i
d = 1 1/2-in. light-gage steel deck	0
e = 10-in. air space (steel joists)	0.85
f = 3/8 in. gyp. lath & 1/2 in. l.w. agg. plaster	0.64
g = Inside surface air film (still)	0.61
	$\Sigma R = 2.60 + R_i$

FIG. 5-8 *Thermal-insulation calculation.*

By Formula (5-1) $U = \dfrac{1}{\Sigma R} = 0.20$

Heat Gain or Loss

Total rate of heat gain or loss Q (in Btu per hour) through a roof is computed by:

$$Q = U \times \text{roof area} \times \text{temperature difference}$$
$$\text{between inside air and outside air} \quad (5\text{-}2)$$

To calculate the temperature at any parallel plane through the roof, use one of the following formulas:

$$T_x = T_i - \frac{\Sigma R_x}{\Sigma R}(T_i - T_o) \qquad \text{for winter conditions} \qquad (5\text{-}3)$$

$$T_x = T_o - \frac{\Sigma R_x}{\Sigma R}(T_o - T_i) \qquad \text{for summer conditions} \qquad (5\text{-}3a)$$

where T_x = temperature at plane X
$\quad T_i$ = inside temperature
$\quad T_o$ = outside *roof surface* temperature (for summer condition)
$\quad \Sigma R_x$ = sum of R values between the warm side and plane X

EXAMPLE: Calculate the deck temperature of the previous example, with $T_o = 150°F$, $T_i = 75°F$.
By Eq. (5-3a), we have

$$T_x = 150 - \frac{0.33 + 2.56}{5.13}(150 - 75)$$
$$= 150 - 42$$
$$= 108°F$$

Without insulation, the deck temperature would be computed as follows:

$$T_x = 150 - \frac{0.33}{2.43} \times 75$$
$$= 150 - 10$$
$$= 140°F$$

Thus, the deck temperature without insulation is 32°F higher than its temperature with insulation.

SIX

*Elements of
the Built-up
Roofing Membrane*

MEMBRANE PRINCIPLES

The built-up roofing membrane is the weatherproofing component of the roof system.   It comprises three basic elements built like a multideck sandwich on the structural deck or insulation.   Alternate layers of felts and bitumen form a flexible roof cover with sufficient strength to resist normal expansion and contraction forces.   A third element, a surfacing normally of mineral aggregate, is embedded in the top surface of bitumen.

The *bitumen* is the waterproofing agent, and thus the most important membrane element.   If it had sufficient fire resistance, strength, rigidity, and weathering durability, the membrane could theoretically (though not practicably) be built entirely of bitumen.

Felts stabilize and reinforce the membrane.   The felt strands form a matrix that restrains the bitumen from flowing in hot weather and resists cracking in cold weather.   The felts also help the mopper apply the bitumen uniformly.

Aggregate surfacing protects the bitumen from damaging infrared and ultraviolet solar rays.   Through a combination of heat and photochemical oxidation, these rays accelerate the embrittlement and cracking of bitumen.   Mineral surfacing provides a fire-resistive skin that prevents flame spread and protects the bitumen from abrasion caused by rain, wind, and occasional foot traffic.   It can help resist the corrosion of acid mists condensing on the roof in industrial areas. An aggregate surfacing even serves as ballast resisting wind uplift and as a shield against the impact of hailstones.

As still another function, the aggregate makes feasible the pouring of a heavy surface flood coat of bitumen, which waterproofs the membrane. The closely massed aggregate forms tiny dams that retard the flow of heated bitumen and allow it to congeal to a depth about three times as thick as a mopped layer (60 lb per square of asphalt versus 20 lb per square of interply mopping).

Uniform, continuous interply moppings form the foundation of a good built-up roofing membrane.   The poured top coat, the external weather-resistant skin, can be corrected if deficient.   But the repair of deficient interply moppings, which can also cause premature membrane failure, is more difficult.

MEMBRANE MATERIALS

Bitumens

Chemically, bitumens are heavy hydrocarbons, limited in the roofing industry to asphalt and coal-tar pitch.   Physically, they are character-ized by low permeability to moisture, good adhesion and cohesion, and slow, continuous deformation under shearing stress.   At high tempera-tures, bitumens deform as viscous fluids; at low temperatures, they deform as elastic solids.

Asphalt is a dark brown to black mixture consisting primarily of carbon and hydrogen, minor amounts of sulfur, and traces of nitrogen, oxygen, and metals.   It is highly viscous at atmospheric temperatures and relatively inert chemically.   Most asphalt used in the United States is produced from the residue of petroleum distillation.   After gasoline, lighter oils, and other volatile products have boiled away, the black, sticky residue from the distillation of crude oils is known as straight-run

FIG. 6-1   *Enlarged cross section through built-up membrane shows laminations of felt and bitumen topped with aggregate surfacing. Note variation in thickness of interply mopping.   (National Bureau of Standards.)*

table 6-1 PROPERTIES OF BUILT-UP ROOFING BITUMENS

| | Asphalt* | | | | | | | | Coal-tar pitch† | |
| | Type I (dead-level) | | Type II | | Type III (steep) | | Type IV (special steep) | | | |
	Min.	Max.	Min.	Max.	Min.	Max.	Min.	Max.	Min.	Max.
Roof slope (in./ft)	0	½‡	½	3	½	6	½	Steep	0	½‡
Softening point§ (°F)	135	150	160	175	180	200	205	225	140	155
Flash point (COC) (°F)	437	...	437	...	437	...	437	...	248	
Penetrations										
@ 32°F, 200 g, 60 sec	3	...	6	...	6	...	6	...		
@ 77°F, 100 g, 5 sec	18	60	18	40	15	35	12	25		
@ 115°F, 50 g, 5 sec	90	180	...	180	...	90	...	75		
Ductility, @ 77°F	10	...	3	...	3	...	1.5	...		
Loss, @ 325°F, 5 hr (%)	...	1	...	1	...	1	...	1	50	
Penetration of residue (% of original)	60	...	60	...	60	...	75	...		
Total bitumen (soluble in CS₂)	99	...	99	...	99	...	99	...		
Specific gravity, 77/77	...	...	...	...	...	...	...	...	122	134
Distillation residue (%)	...	...	...	...	...	...	...	...	90	

* Per ASTM D312-64.
† Per ASTM D450.
‡ Steeper slopes may be permitted with adequate nailing or mechanical fasteners; see manufacturer's recommendations.
§ Asphalt softening point per ASTM D2398; coal-tar pitch softening point per ASTM 61.

asphalt. Further refining, by blowing air or other gases through the heated mass, produces harder, less sticky material called blown asphalt.

Mopping asphalts come in a wide range of viscosities and softening points refined to a range of 135 to 215°F for use on various slopes. Asphalts in the lower softening-point range (135 to 150°F) are known as "dead-level" asphalts; more viscous asphalts, known as "steep" asphalts, are suitable for roofs of greater slope. Special steep asphalt (205 to 215°F softening point) is used for roof slopes up to 6 in. per ft. (see Table 6-1).

Coal-tar pitch, a by-product of the destructive distillation of bituminous coal in the manufacture of coke or gas, is a denser bitumen, distilled to a softening point of 140 to 155°F. The distribution of molecules and lack of strong intermolecular attractions give coal-tar pitch excellent cold-flow, self-healing properties, through which the pitch flows together and closes cracks formed at cold temperatures. The cold-flow property limits its use to roof slopes of 1 in. per ft or less.

Overheating bitumen shortens its life. Asphalt overheated by 50, 100, and 150°F for 4-hr periods becomes progressively more brittle; it checks and cracks in weatherometer tests in proportion to the degree of overheating. It also loses adhesive strength. Overheated coal tar pitch suffers similarly deleterious effects. In some cases, asphalt may become softer with prolonged overheating (softening-point fallback).

Asphalt and coal-tar pitch are incompatible and, as a rule, should not be used in combination. One bitumen softens as it absorbs the exudate from the other, which simultaneously hardens. The worst mixing occurs when asphalt melts in a kettle coated with the remnants of coal-tar pitch (or vice versa). Such mixing produces a mongrel bitumen with intolerably unpredictable properties.

There are, however, exceptions to the general restriction. Asphalt flashing materials and asphalt-coated base sheets (with top side of stabilized asphalt coating and mineral surfacing) can be used with coal-tar pitch.

Felts

Roofing felts are nonwoven fabrics classified as either organic or inorganic (asbestos or glass fiber). The paper manufacturing process used to produce organic and asbestos felts orients their fibers in the longitudinal (machine or roll) direction, making them stronger in that direction than across their width.

Organic felts are manufactured chiefly from combinations of felted papers and shredded wood fibers. (Organic felts, sometimes erroneously called rag felts, have little rag content.) Organic felts are saturated in coal tar pitch or asphalt. Asphalt-saturated organic felt costs least and is by far the most popular roofing felt on the market.

Asbestos felts are manufactured from asbestos mineral—threadlike fibers of magnesium and calcium silicate. They resist the deterioration from humidity, solar radiation, and other destructive agents better than organic felts.

Glass-fiber felts are flexible asphalt–glass-fiber mats designed for use with asphalt in built-up roofing assemblies. When lightly coated with asphalt, glass-fiber felt gains strength.

Felts are manufactured into three basic kinds of sheets: saturated felts, coated sheets, and mineral-surfaced sheets.

Saturated felts have been saturated or impregnated with bitumen.

Coated sheets, which are highly impermeable to moisture, are first saturated and then coated with a factory-applied film of asphalt, normally on both sides.

Mineral-surfaced sheets are coated sheets with an added step in the manufacturing process—the embedment of mineral granules in the exposed surface.

Saturated felts are used as ply felts and occasionally as base sheets. Coated felts are used chiefly as base sheets, sometimes as ply felts. Mineral-coated felts are used as surfacing (cap) sheets.

All felts come in 36-in.-wide rolls.

Ply Felts   Ply felts come in several weights, ranging from 7½ lb per square (Type 8) for the lightest glass-fiber felt to 28 lb per square (Type 30) for asbestos and organic felt (see Table 6-2).

Asphalt-saturated organic and asbestos ply felts for built-up roofing are generally manufactured with tiny, closely spaced perforations designed to permit entrapped air and moisture to escape during application. The porous glass-fiber felts do not require perforation. As the broom smooths out the felt, the perforations (or openings in the glass-fiber felt) facilitate escape of air or water vapor, which if not released could ultimately blister the membrane.

Felts saturated with coal-tar pitch should never be used with asphalt, and asphalt-saturated felts should not be used with coal-tar pitch. Such mixing has largely unpredictable, but usually harmful, effects on the bitumen softening point.

Surfacing

For hot-mopped, built-up roofing membranes, there are three basic classes of surfacing:

- Mineral aggregate (embedded in bitumen)
- Asphalt (hot- or cold-applied)
- Mineral-surfaced cap sheet

Mineral-aggregate surfacing offers the degree of protection needed for low-sloped roofs (under ½ in. per ft). The thick flood coat needed for

table 6-2 TYPICAL BUILT-UP ROOFING FELTS

Designation	ASTM Specification	Type	Minimum weight, lb/sq	Minimum weight of saturant, lb/sq	Minimum tensile strength, lb/in.	
					Machine	Cross
Organic, asphalt...	D226-60	15	13	7.3	30	15
		30	26	15.0	40	20
Organic, coal-tar pitch.............	D227-56	15	13	7.3	30	15
Asbestos, asphalt...	D250-68	15	13	3.6	20	10
		30	28	7.0	40	20
Glass fiber, asphalt.	D2178-63T	8	7.5*	4.5	17	12
		15	14*	9.0	19	12
Coated asbestos, asphalt...........	D655-47 (1965)	50	50	8.8		
Coated organic, asphalt...........	D2626-67T	I	35	7.3	35	20
		II	37	9.8	45	20

* Average of all rolls.

the top layer of waterproofing bitumen is attained only through the use of aggregate.   A sound, opaque aggregate is needed for protection.

Aggregate surfacing generally consists of gravel, slag, or crushed rock set in a flood coat of bitumen.   The maximum 3-in. slope limit generally recommended by roofing manufacturers stems from two construction problems:

- Placing a 60- to 75-lb flood coat that will stay in place until it has set
- Keeping gravel from sliding down the roof slope in hot weather

Aggregates must be clean, dry, and opaque.   They must be graded to sink into the bituminous flood coat and to nest properly, i.e., to fill the voids between the individual particles.   (ASTM Specification D1863 calls for a minimum size of $\frac{1}{4}$ in., maximum $\frac{5}{8}$ in.)   They must be hard, but not brittle, to resist abrasion from rain, wind, and occasional foot traffic.

The most common roof surface aggregates are river-washed gravel, crushed stone, and blast-furnace slag—a fused, porous substance separated in the reduction of iron ores.

Dolomite (marble chips) can give a roof a good heat-reflective surface, but it has several drawbacks.   Small translucent chips admit damaging solar radiation, which may penetrate to the flood-coat bitumen.   The

FIG. 6-2   *The flood coat of an aggregate-surfaced roof is poured (manually as above, or mechanically), not mopped, like the lower layers of bitumen.   (GAF Corporation.)*

resulting embrittlement of the bitumen can loosen the aggregate, cause premature weathering, and thus shorten membrane life.   Moreover, dolomite chips are often coated with dust, which also weakens the aggregate bond to the bitumen.

Other roof-surfacing aggregates include colored rock, crushed tile, brick, limestone, and volcanic and asbestos rock.   These aggregates are generally more expensive than gravel and harder to control.   Before deciding on these more exotic surfacing materials, the designer should consult with a roofing contractor or experienced bitumen manufacturer.

For slopes of 1 in. per ft or more, where water runoff is rapid and there is no threat of ponded water even for short intervals, a smooth-surfaced roof with an asphalt-mopped asbestos felt or a mineral cap-sheet surfacing is permissible.   A smooth-surfaced roof relies largely on the water repellence of inorganic felt, which, unlike organic felt, does not absorb moisture.

A *smooth-surfaced roof* is topped either with 15 to 25 lb (per square) of hot-applied asphalt mopping or cold-applied cutback or emulsion applied to an asphalt-saturated inorganic felt at a spreading rate of 25 to 50 sq ft per gal.   Organic felts, occasionally used for smooth-surfaced roofs, are not recommended.   For suitably sloped roofs, the designer may choose a smooth-surfaced membrane to reduce roof dead-load (by about 4 psf), to facilitate the discovery and repair of membrane splits or cracks, or merely for the aesthetic value of an ungraveled surface.

Actually, the felt functions as the surfacing of a smooth-surfaced roof, despite its normal protective asphalt coating.   According to some experts, the bared inorganic felt is better left exposed after the original protective asphalt mopping has oxidized and eroded.   Remopping may thicken the exposed asphalt surface irregularly, thereby increasing the risk of alligatoring.   Alligator cracks tend to trap water, which eventually penetrates the plies.   In contrast, the smooth surface of an exposed, suitably sloped inorganic felt surface allows fast runoff without trapping water.   There are, in fact, smooth-surfaced roofs in which the top felt is a factory-coated, asphalt-saturated inorganic felt, with no further field coating.

A *mineral-surfaced cap sheet* (of organic felt) is also suited for roofs with slopes of 1 in. per ft or more.   For the steeper slopes where aggregate surfacing tends to slide, the mineral-surfaced cap sheet provides a stable surfacing.   A common specification calls for a cap sheet over two 15-lb felts.

Although organic cap sheets are limited to a 1-in. minimum roof slope, inorganic cap sheets are used for slopes as low as $\frac{1}{4}$ in. per ft.   They are, however, more vulnerable to standing water than aggregate-surfaced roofs.

Although smooth-surfaced roofs with inorganic felts have performed satisfactorily, aggregate surfacing is generally preferable to smooth or cap-sheet surfacing for roofs of low slope.   Aggregate surfacing offers superior weather resistance as well as superior resistance to fire and hailstone impact.   It also offers superior wind-uplift resistance, because its rough surface tends to destroy the smooth laminar air flow that increases negative wind-uplift pressures and its slightly greater roof weight acts as ballast.   With little maintenance, an aggregate-surfaced roof may last 20 years or more, much longer than a smooth-surfaced membrane similarly neglected.   (An asphalt surfacing should not be expected to last more than 3 to 6 years before it requires preventive maintenance or renewal.)   Material storage, traffic, or other abuse can, however, greatly shorten an aggregate-surfaced membrane's life by forcing aggregate through the flood coat and into the felt plies.

A smooth-surfaced roof is superior to an aggregate-surfaced roof in one respect.   Because the aggregate may hide a developing blister or split, it is generally more difficult to detect defects in an aggregate-surfaced membrane than it is in a smooth-surfaced or cap-sheeted membrane. Repairing or replacing an aggregate-surfaced membrane is more costly; it often requires the laborious job of scraping the flood coat.

Reflective Surfacings

Reflective surfacings can help protect the bitumens from the deterioration caused by photo-oxidative chemical reactions that embrittle bitumens. Color plays the key role in reflective cooling.   Under a sunny summer sky, with air temperature 95°F, a black roof surface can climb to 180°F, whereas a white roof might be 50°F cooler.   A gray surface would produce intermediate temperatures (see Table 6-3).

table 6-3 INSULATED ROOF
SURFACE TEMPERATURES
FOR DIFFERENT COLORS

Membrane color‡	Surface temperature, °F	
	Summer*	Winter†
Black	190	−20
Dark gray	175	−20
Light gray	160	−20
White	140	−20

* Air temperature 95°F on sunny day.
† Clear night of 0°F, when terrestrial heat
radiation is unreflected by clouds.
‡ As color changes with aging, dirt accumula-
tion, etc., surface temperature changes.

Reflective surfacings are available for all three classes of roof surfacings. Reflective aggregates include white marble chips and other light-
colored aggregates. Reflective cap sheets are surfaced with light-mineral
granules. For smooth-surfaced roofs, aluminum or other light, pig-
mented coatings can be brushed, rolled, or sprayed. Such coatings must
be renewed every few years.

THE BASIC MEMBRANE
SPECIFICATION

Evolution of Built-up Membrane
Specifications

The wide variety of roofing specifications, with different numbers of felt
plies and moppings, has evolved from two basic ancestors. The older
is the five-ply wood-deck specification, in which the two lower plies are
nailed to the deck and the top three mopped. The younger specification, for concrete decks, in which the base plies are mopped to the deck,
has one fewer ply (four) and one more bituminous mopping (four).

Subsequent modifications have reduced the required number of plies
for specially designed systems to two, for a minimum built-up roofing
membrane specification. Most built-up roofs today are designed for a
minimum 20-year life expectancy. In selecting a specification for 10,
15, or 20 years, the designer manifests his judgment on the relative value
of durability versus economy or capital cost versus maintenance cost
(see Table 6-4).

table 6-4 TYPICAL BUILT-UP MEMBRANE SPECIFICATIONS

	20-year manufacturer's bond							15-year manufacturer's bond				
Type[a]	A-O	A-O	A-S	A-G	A-G	A-OC	P-O	A-O	A-OC	A-S	A-G	P-O
Slope limits[b] (in./ft)	0-1	0-1	0-6	0-6	0-3	¼-3	0-1	0-3	0-3	0-6	0-3	0-1
Base sheet (lb/sq)	43	43	43	...	...	...	43	43	43	43	...	43
No. felt plies (in addition to base sheet, if used)	3	3	3	3	2	2	3	2	2[f]	2	3	2
Interply mopping weight (lb/sq)	20	20	20	25-32	32	20	25	20	20	20	32	25
Flood coat (lb/sq) (only for aggregate surfacing)	55-60	55-60	55-60	...	60	60	65-75	55-60	...	55-60	...	65-75
Surfacing:												
Aggregate[c]												
Smooth-surfaced[d]												
Mineral-surfaced roll roofing (lb/sq)	...	...	...	72	...	90[e]	...	...	[f]	...	...	...

[a] Key to type: first letter denotes bitumen (A = asphalt; P = coal-tar pitch). Second letter denotes felt type (O = organic; S = asbestos; G = glass-fiber; OC = organic coated).

[b] Minimum recommended slope for any roof = ¼ in. per ft.

[c] Aggregate surfacing is either (1) slag @ 300 lb/square, or (2) gravel @ 400 lb/square.

[d] Smooth-surfacing is either (1) hot-mopped asphalt (20 to 25 lb/square); (2) clay-type asphalt emulsion (3 gal/square); or (3) asphalt roof coating (2 to 4 gal/square).

[e] Mopped with hot asphalt (20 lb/square).

[f] Felts are wide-selvage roll roofing with 19-in. side lap.

Field Manufacturing the Membrane

The typical modern built-up roofing specification calls for a coated base sheet nailed or mopped to the deck. Following the application of the base sheet come several plies of felt (from two to four), which are *shingled* (see Fig. 6-3).

Specification Roof

Decks: Structural wood fiber, poured gypsum, lightweight insulating concrete

Max. slope: 1 in. per ft

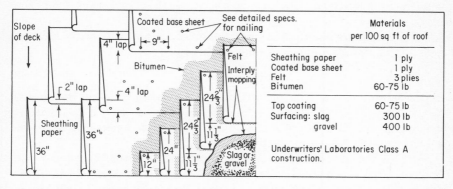

Materials per 100 sq ft of roof	
Sheathing paper	1 ply
Coated base sheet	1 ply
Felt	3 plies
Bitumen	60-75 lb
Top coating	60-75 lb
Surfacing: slag	300 lb
gravel	400 lb
Underwriters' Laboratories Class A construction.	

FIG. 6-3 *Shingling of roofing felts requires their simultaneous application, as shown above. With three shingled plies, the exposure of each 36-in.-wide felt is 11⅓ in., computed by dividing the felt width minus 2 in. by the number of plies. Diagram (top) shows a four-ply roof (coated base sheet plus three plies of saturated felts, the felts as shown in the photograph).*

Application of the felt starts at an edge, or low point (so the shingled joints will not "buck" water) with the cutting of a felt into strips whose widths are an even factor of the 36-in. felt width divided by the number of plied felts. Thus for three-plied shingled felts (a four-ply membrane, including base sheet), the edge strips would be 12, 24, and 36 in.

Next come the regular 36-in.-wide shingled felts. The overlap dimension is computed by dividing the felt width minus 2 by the number of shingled plies. Thus for three shingled plies, the overlap $= (36 - 2)/3 = 11\frac{1}{3}$ in., which is also the distance of the first sheet from the edge.

This overlap assures that any vertical cross section will always have at least the minimum number of required plies.

Theory of Shingling

Practices used in the field manufacture of a built-up roofing membrane blend common sense, tradition, and scientific analysis. The practices themselves are fairly standardized, but there is disagreement within the industry on the reason for certain practices.

Shingling of interply felts, though universal, actually reduces the waterproofing quality of the membrane. If the plies were laid in alternating, plane laminations, with an entire ply placed and mopped before the application of the next ply, the planes of mopped bitumen would provide better secondary waterproofing defense than the moppings between the shingled plies. Shingled plies can admit a slow passage of water down into the bottom of the membrane.

Foundation, wall, and concrete-topped basement roofs are waterproofed with alternating planes of felt and bitumen. Unlike these other waterproofing laminations, however, a built-up roofing membrane is exposed to the weather, and this is the basic reason for shingling the felts. Whereas the waterproofing sandwiched between two concrete slabs requires only slight compressive strength, the built-up membrane must resist horizontal shearing forces and wind-uplift stresses tending to delaminate it. Shingling the felts increases membrane strength by eliminating potentially weak slippage planes of bitumen sandwiched between planes of felt.

Coated Base Sheet

The *coated base sheet* slows the passage of water vapor entering the membrane from the underside. As a highly impermeable film on the underside of the membrane, the base sheet is applied as a single ply, not shingled with the ply felts.

Use of the coated base sheet has become an industry standard for organic and asbestos felt, although not for glass-fiber felts. In the basic nailable wood-deck specification, one coated base sheet replaces two saturated felt plies. In the further evolution of this process, complete two-ply coated felt membranes, in which the two plies of coated felt replace a coated base sheet and three plies of saturated felt, have entered the market.

Application of a base sheet over a poured gypsum or lightweight insulating concrete substrate poses special problems because these poured materials have high moisture content. Current methods include nailing and hot mopping, but sometimes conditions justify placing a layer of board insulation on the concrete surface to improve the substrate. Application of the membrane over these moisture-bearing poured substrates is always sensitive.

For nailing, the requirement of 40-lb pullout strength per nail limits this anchorage method. For lightweight insulating concrete substrates to be nailed, a minimum 1:6 cement-aggregate ratio (by volume) for perlite and 1:4 for vermiculite is considered satisfactory. In any case, the designer should investigate a fastener for minimum 40-lb pullout strength. Some types of special fasteners provide better pullout resistance than standard roofing nails.

Because of the difficulty of satisfying the nail pullout strength requirement, hot-mopping is more widely used than nailing over lightweight insulating concrete. Solid mopping provides greater adhesive strength and greater resistance to vapor migration into the felts. As disadvantages, however, solid mopping may produce stress concentrations, thus increasing the chances of membrane splitting. Intermittent mopping— spot or channel—reduces the threat of stress concentrations, but increases the threat of water-vapor migration and introduces the threat of blistering, resulting from the inadvertent closing of vapor escape paths if the mopper drips bitumen into the open spaces of the intermittent mopping pattern.

In a more conservative approach for applying a base sheet to these poured substrates, the designer adds a thin layer of preformed, vapor-porous insulation (e.g., fiberglass) over a coated sheet, to provide an improved substrate. The coated sheet should be nailed over gypsum, but solid mopped over lightweight insulating concrete.

Temporary Roofs

A temporary roof, comprising built-up layers of felt or a heavy, coated base sheet, is advisable when occupancy is critical or when the job is hampered with any of the following conditions:

- Prolonged rainy, snowy, or cold weather
- Necessity of storing building materials on the roof deck
- Mandatory "in-the-dry" work within the building before weather permits safe application of permanent roofing system
- Large volume of work on roof deck by trades other than roofer

A minimum thickness of insulation must be applied over fluted steel decks as a substrate for the temporary roof.

All temporary materials, including insulation, should be removed before installation of the permanent roofing.   The risk of damaging the temporary roof, subjected to a heavy volume of construction traffic, is too great to warrant its use as a vapor barrier or bottom plies of the permanent roof.   Owners who save the costs of replacing temporary roofing materials are gambling small current gains against potentially large future losses.

Phased Application

In phased application, the felts are applied in two (or more) operations, with a delay between operations.   At the break in felting application, the felts are necessarily unshingled.   This break may occur, for example, between application of a coated base sheet and the three shingled ply felts of a four-ply membrane.

Phased application offers conveniences to a roofing contractor; while the uncompleted roof serves as a temporary roof, he can spread his workmen around many projects on good working days.   It can, however, create future roofing problems for architect, roofing contractor, and owner.

Leaving saturated (uncoated) organic felts exposed, even overnight, can result in moisture expansion of the felts.   Subsequent drying will shrink the felts.   It may build in contraction stresses and ultimately split the membrane, as entrapped moisture evaporates and expands.   Moisture absorbed in the felts can curl their edges, as the exposed side dries and contracts (see Fig. 6-4).   These curled edges are difficult to flatten. They may protrude through the flood coat and aggregate surfacing, where they remain as water conduits into the membrane.   Felts left unsurfaced for any appreciable time accumulate dust that can weaken the bond with the bituminous flood coat.   Uncompleted membranes are more vulnerable to traffic damage than completed, surfaced membranes.

Leaving even a coated base sheet exposed for long time intervals as a temporary roof is bad roofing practice.   Though far less absorptive and

FIG. 6-4   *Saturated (uncoated) felts left exposed even for short intervals absorb moisture and then curl at the edges as the top side dries and contracts.   (Koppers Co.)*

much more weather-resistant than saturated felts, coated felts are far from impervious to prolonged exposure.   If exposed for a week or more to the alternating cycles of normal atmospheric evaporation and condensation, they will start to absorb moisture.   After prolonged exposure even an apparently dry coated felt may be a moisture-laden source of future membrane problems.

To avert these problems with exposed felts, the roofer often applies a light bituminous mopping (glaze coat) to protect the felts until they get their final surfacing.   This expedient helps in some ways, but may present an easily damaged traffic surface, one that is tacky in summer and brittle in winter (see Fig. 6-5).

In summary, phased application with uncoated felts left exposed even overnight should be prohibited.   Leaving coated felts exposed for more than short intervals—a day or two at most—should also be prohibited.

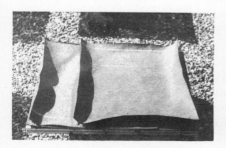

FIG. 6-5   *Glaze coating, though not a perfect remedy, does maintain a plane surface for uncompleted membranes.   (National Bureau of Standards.)*

COLD-PROCESS MEMBRANES

Cold-process bituminous roofing replaces hot-mopped bitumens with cut-back bitumens applied at ambient temperature, thus eliminating kettles and fuel from the job site. Cold-process bitumen was formerly used chiefly for maintenance and repair work, for which the small quantities of required bitumen justified the much higher unit cost of the cold-process bitumen. Cold-process bitumen may be used for other reasons:

- Air pollution regulations of some United States cities (notably Los Angeles) ban open-air heating of bitumen under certain smog conditions.
- Fire safety regulations may require the presence of firemen at the kettle heating in some jurisdictions.
- Hoisting a kettle to a skyscraper roof for a relatively small area may be uneconomical.
- In outlying areas, the cost and trouble of transporting the fuel and kettle may outweigh the normal economies of hot bitumen.
- Maintenance personnel unfamiliar with built-up roofing can make repairs with cold-process bitumen.

Cold-process asphalt emulsions may meet the required Class A or B fire classification for a smooth-surfaced membrane that would not qualify with hot-mopped asphalt surfacing [see Chap. 8, "Fire Resistance" (Roof Covering Ratings), pp. 123–125].

Cold-process bitumens are brushed or sprayed on the surface. They are refined to liquid state either through a solvent (cutback) or through emulsification (see Fig. 6-6).

The solvent-thinned cutbacks solidify in place as the organic solvent evaporates. The bituminous emulsions solidify as water evaporates, leaving the bitumen and emulsifying agent as a residue.

Asphalt-based emulsions come in two types. One type has a mineral colloid (bentonite-clay) emulsifying agent; the other type has a chemical emulsifying agent. The clay type resists weathering erosion better than the chemical type; therefore it is recommended for use in roofing applications.

Asphalt emulsions serve only as surfacing bitumen. Cutbacks serve as interply cement and, less frequently, as surfacing bitumen. As a surfacing, cutbacks are inferior to asphalt emulsions.

Cold-process membranes are limited to smooth-surfaced roofs (often with reflective coating) or mineral-surfaced cap sheets. Since it is impractical to get the required depth of a flood coat, cold-process bitumens are not adaptable to aggregate surfacing. They should be sloped at least $\frac{1}{4}$ in. per ft.

FIG. 6-6   *Clay-type asphalt emulsion surfacing for cold-process membrane is either brushed (as above) or sprayed onto coated felts.*

Design and application of a good cold-process membrane are somewhat more critical than for a hot-applied membrane, and several practices are consequently altered. Because they fuse more completely with cold cements than saturated felts, coated felts should be used in cold-process membranes. To achieve good interply bond, these felts should be unrolled prior to application. They are precut into 18-ft strips and laid on the roof to flatten for several hours before application (see Fig. 6-7). Careful brooming is mandatory to get proper sealing of side and end laps,

FIG. 6-7   *Felts for cold-process bituminous membrane are laid dry on deck to flatten, to assure better adhesion.* (Southwestern Petroleum Co.)

FIG. 6-8   *Rolling the longitudinal side laps of a cold-process membrane is recommended by some manufacturers to assure good interply bond.*

which are greatly increased by the felt cutting.   Some manufacturers recommend rolling a cold-process membrane with a weighted roller at the end of the day's work, with special attention to the seams to eliminate fishmouths and assure good interply bond (see Fig. 6-8).

PREMATURE MEMBRANE FAILURES

There are three major modes of premature built-up membrane failure:

- Wrinkling (ridging or buckling), culminating in cracking
- Splitting
- Blistering

Less widespread, but potentially no less troublesome, are:

- Slippage
- Delamination
- Alligatoring
- Surface erosion

Wrinkle Cracking

Wrinkling, the most common kind of membrane failure, became a major problem with the inclusion of insulation between deck and membrane. Wrinkles generally appear in parallel lines directly above continuous, longitudinal insulation joints. Less frequently, wrinkles also appear above the staggered, transverse insulation joints, along with the longitudinal wrinkles. They may start out narrow and low, remaining unnoticed until erosion of aggregate surfacing exposes the bitumen. Left alone, they can ultimately grow to a 2-in. height (see Fig. 6-9).

Wrinkling failure may occur when high surface temperature causes the fluid bitumen to flow down the ridge slopes. Repeated bending at the ridge, caused by cyclic elongation and contraction accompanying alternate wetting and drying of the felts, ultimately cracks the membrane at the ridges.

Wet felts cause roof wrinkles. Moisture may come from installation of wet materials, condensation of infiltrating water vapor, or a roof leak. Moisture migrating upward from wet insulation is also a major cause of wrinkling. Stacking insulation boards outdoors in wet weather, even if covered by tarpaulins, can cause heavy moisture absorption, concentrated in the board edges. The wet edges fail to bond to the membrane. Left free to move and open the joints, further weakening their bond to the membrane, these wet edges also supply water to the felt above. For a roof so constructed, wrinkling is almost inevitable.

Wrinkles form with the quick heating of a membrane when there are one or more wet felts. The heat-softened bitumen permits the wet felt to swell and expand the membrane. Higher surface temperatures promoted by the insulation itself aggravate membrane expansion. The expanding membrane wrinkles over the insulation joint, where its bond to the substrate is weakest. The wrinkle-forming action recurs through the spring and early summer when moisture moves up at night and

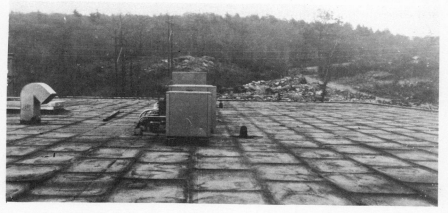

FIG. 6-9   *Wrinkling normally occurs over the joints of dimensionally unstable insulation boards.*

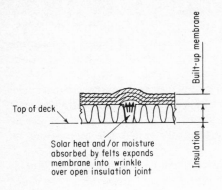

FIG. 6-10   *Wrinkling mechanics.*

evaporates in the heat of the following day.   Because the membrane stiffens at low temperatures, it will never contract to its original length. Thus even after drying, a membrane wrinkle can be permanent.

The widespread practice of unrolling felts *parallel* to the continuous longitudinal insulation joints promotes wrinkling. Since wet felts expand much more in the transverse than in the longitudinal direction, the normal practice of aligning the felt rolls parallel with the longitudinal axis of the board insulation maximizes wrinkling, since the greatest substrate movement occurs at these longitudinal joints (see Fig. 6-10). Organic felts expand and contract much more than inorganic felts, and thus wrinkle more easily.

Thick insulation also heightens the risk of wrinkling.   Its joints are normally wider and freer to expand and contract than thinner insulation. A roof membrane placed over a 2-in.-thick insulation tends to wrinkle more than the same membrane over two 1-in.-thick layers, with staggered and consequently narrower joints.

Blisters

Blisters can range in size from tiny, virtually undetectable spongy spots to bloated spaces of 1-ft-high, 50-sq-ft area.   A small blister may result from air or water vapor entrapped *between* the plies.   It is less serious than a big blister, which indicates a bond break between the entire membrane and the substrate (see Fig. 6-11).

High temperatures that evaporate water and increase vapor pressures produce blisters.   Under the 160°F temperature frequently reached on roof surfaces, water will expand 1,500 times after evaporating.   Confined within a constant volume, water vapor raised 100°F will exert a pressure of 600 psf.

A blister cannot grow without limited leakage of a roofing sandwich containing porous insulation.   Unlike a sealed chamber, an incipient

FIG. 6-11   *Blisters indicate heated water vapor entrapped between substrate and membrane, under pressures up to* 300 *to* 400 *psf.* (*Koppers Co.*)

blister without perfect closure will suck air and moisture at night when the temperature drops and the cracks in the bitumen open. During the day, solar heat will expand the blister, but the cracks will close and heal as the bitumen liquefies. Normal temperature cycling thus produces a cyclic pumping action that enlarges the blister. The area under pressure increases proportionately with the square of the blister diameter, while the peripheral bond resistance between membrane and substrate increases linearly, thus reducing the blister's resistance to further growth (see Fig. 6-12).

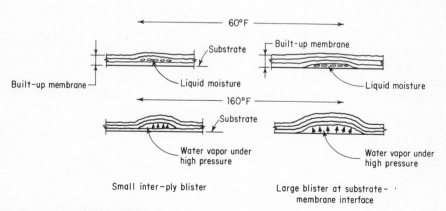

FIG. 6-12   *Blistering mechanics.*

Large roofs with vapor barriers are especially vulnerable to blistering. For a given roof plan shape with the same edge-vent spacing, the volume of air to be released per foot of perimeter doubles as the perimeter doubles. Thus on a large roof, with longer restricted paths from central area to edge, the blister-forming internal pressure under rising temperatures is much greater than on a small roof. (Stack venting can alleviate this condition.)

Splitting

Since it immediately admits water, splitting is the most serious of the three major premature failure modes. It is common in cold climates, especially in insulated roofs. Built-up roofing membranes generally split parallel to the longitudinal (rolled or machine) direction of the felts, thus indicating their weak transverse direction as well as their high thermal movement in this direction. Like wrinkles, splits generally occur directly above and along the continuous longitudinal joints of board insulation (see Fig. 6-13).

Splitting has many causes, acting singly or in combination. Among these are movement of the substrate accompanying thermal contraction, insulation slippage over the deck, deck deflection, or change of deck span. Two other factors in splitting—thermal movement of the membrane and drying shrinkage in the felts—must act in combination with at least one other factor to develop sufficient tensile stress to split a well-anchored membrane.

Splitting often occurs at subfreezing temperatures. When the insulation has contracted either through drying or falling temperature, the opened joints subject the membrane to stress concentrations. A sudden drop in temperature may raise the tensile stress enough to break the membrane.

The presence of ice on a ponded roof will increase the membrane's tendency to split at low temperatures. Below 32°F ice contracts with a thermal coefficient $4\frac{1}{2}$ times that of steel. If it is keyed into the aggregate surfacing, it can split the membrane when, under a sharp temperature drop, it cracks like ice on a frozen lake.

Splitting can also result from solid-mopping a base sheet to a gypsum deck. Gypsum decks often develop shrinkage cracks over the bulb tee

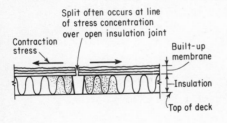

Split often occurs at line of stress concentration over open insulation joint

Contraction stress

Built-up membrane

Insulation

Top of deck

FIG. 6-13 *Splitting mechanics.*

subpurlins, where the thinned section is weakest. If the base sheet is nailed, it can accommodate the small movement associated with the shrinkage crack within the minimal 9-in. nail spacing usually specified. But if the base sheet is solid-mopped, the local strain over the crack may be 20 to 30 times greater than the local strain in a nailed sheet. In cold weather, splitting of the membrane over a solid-mopped base sheet becomes even more probable.

Excessive surface mopping can increase the hazard of splitting by increasing the contraction stresses as the membrane shrinks in cold weather. (The thickened bitumen increases the contraction stress in the felts.)

The danger of membrane splitting, as well as wrinkling, can be reduced by unrolling felts *perpendicular* to the longitudinal joints of a board insulation substrate (see Fig. 5-2). Since the top edges of the insulation boards along these joints are the most mobile part of the substrate, expansion and contraction of the insulation produce maximum movement at the longitudinal joints, which become lines of stress concentration (see Fig. 6-13). Since the felts are weakest in their transverse direction, the widespread practice of unrolling felts parallel to the longitudinal joint combines maximum stress with minimum strength, thus heightening the risk of splitting.

Membrane Slippage

Membrane slippage, a problem chiefly in sloped roofs, is relative lateral movement between adjacent plies. It often results in a randomly wrinkled appearance, much like a badly laid carpet (see Fig. 6-14).

Slippage can ultimately expose the base sheet, thus thinning the membrane, removing the surfacing, and exposing the remaining weakened portions of the membrane to weathering and other destructive forces.

Slippage results from a complex of factors, usually one or more of the following:

- Phased application (which by eliminating the felt shingling at one horizontal interply joint weakens horizontal shearing resistance)
- Insulated roof membranes
- Inadequate mechanical anchorage
- Bitumen of too-low softening point flowing at high temperature (a possible result of softening-point "fallback" due to overheating)
- Overweight bitumen moppings
- Overweight aggregate surfacing
- High roof surface temperatures

Slippage resulting from low-softening-point asphalt has become so common that several major roofing manufacturers now recommend steep

FIG. 6-14   *Membrane slippage, a problem chiefly in sloped roofs, is
caused by one or more of a complex of factors.   (National Bureau
of Standards.)*

asphalt (ASTM D312, Types III or IV) for interply mopping on all roofs,
regardless of slope.   In the southern states, roughly those south of the
37°N parallel, most manufacturers recommend Type IV asphalt for interply moppings on all roofs.

Delamination

A less common membrane failure, delamination, generally indicates at
least one of the following:

- Insufficient bitumen between felts
- Improper embedment of felts due to inadequate brooming or under-
weight mopping
- Application of bitumen at too-low temperature

FIG. 6-15 *Advanced stage of alligatoring shows exposed felt deteriorating under attack of sunlight and moisture.*

■ Mixing coal-tar-saturated felts with asphalt, or asphalt-saturated felts with coal-tar pitch

■ Water absorption during the winter, followed by evaporation in spring or summer

A delaminated membrane may wrinkle and later crack, thus behaving like a wrinkled membrane.

Alligatoring

Alligatoring occurs only in exposed asphalt, especially steep asphalt. It occurs in smooth-surfaced roofs and sometimes in bare spots of aggregate-surfaced membranes. It consists of deep shrinkage cracks, progressing from the surface down, a result of continued oxidation, aging, and embrittling (see Fig. 6-15).

The alligator cracks can retain water, which threatens eventually to penetrate through to the felts where it can work its familiar mischief.

Alligatoring is aggravated by thick moppings. Recoating an already alligatored surface may make things worse.

Surface Erosion

The displacement of surface aggregate (and sometimes the flood-coat bitumen) often issues in progressive failure. It may be followed by alligatoring, felt deterioration, blistering, or other failures caused by the infiltration of water into the membrane or the insulation.

Application of wet, dirty, or ice-crusted aggregates makes aggregate surfacing vulnerable to the action of the various forces that cause surface erosion. These include:

■ Wind suction, which sometimes reaches negative pressures of 40 or 50 psf behind parapets at building corners.

- Water flow or drip, especially where conductors from a roof above discharge onto a lower roof.
- Rooftop traffic, especially on regular paths or where snow is shoveled, and mechanical installation or repair work is done.

Surface erosion is far less common on smooth-surfaced roofs, where it entails loss of surfacing asphalt, than on aggregate-surfaced roofs. Its major causes:

- Excessive foot traffic
- Weathering
- Discharge of corrosive or solvent-type fumes or liquids on the roof surface

ALERTS

General

1. Check with the manufacturers of built-up roofing and composition flashing to make sure that proposed roofing system and materials are compatible with roof deck, vapor barrier, and insulation.

2. On major roofing projects, require either inspection by the manufacturer or inspection service provided by the owner.

Design

1. Use an asphalt-coated base sheet for first ply in built-up roofing membranes.

2. Do not specify coal-tar pitch on roofs with slope exceeding $\frac{1}{2}$ in. per ft.

3. Specify asphalt with lowest possible softening point suited to slope and climate as follows:

Slope,[1] in. per ft	ASTM designation	Softening point, °F
$\frac{1}{2}$ in. or less	D312, Type I, dead level	135–150
$\frac{1}{2}$–3	D312, Type II, flat	160–170
$\frac{1}{2}$–6	D312, Type III, steep	180–200
3–6	D312, Type IV, special steep	205–215

[1] Warm climates may limit the slopes suitable for asphalts of lower softening point to less than maximum slopes specified above.

4. When hot-mopped bitumen is used to bond a base sheet to insulation, specify ASTM Type III or Type IV asphalt.

5. Apply felts *perpendicular* to the longitudinal, continuous joints of board insulation.

6. Provide trafficways or complete traffic surfacing for roofs subjected to more than occasional foot traffic (see Figs. 6-16 and 6-17). (Mineral

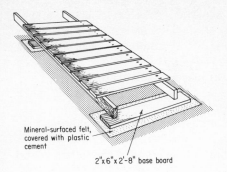

FIG. 6-16 *Construction of wood walkway starts with embedment of mineral-surfaced, coated felt strips in hot bitumen. Next comes a layer of plastic cement, into which the base boards are set. Aggregate surfacing follows, between and around the walkway bases. (Owens-Corning Fiberglas Corp.)*

Mineral-surfaced felt, covered with plastic cement

2"x 6"x 2'-8" base board

aggregate will not protect the membrane against workmen installing mechanical or other equipment over roofing.)

7. In roofs over cold-storage space, do not place membrane in direct contact with cold-storage insulation. (For discussion of cold-storage insulation, see Chap. 5, "Thermal Insulation.")

8. Avoid contact between built-up membrane and metal flashings and other metal accessories, especially below the level where water collects.

Field

1. Prohibit storage of felts over new concrete floors. Require pallets covered with kraft paper. Stack felts on ends. Avoid prolonged storage of felts on site.

FIG. 6-17 *Granule-coated treads, made with an asphalt filler sandwiched between coated felts, are placed in flood coat to form walkway. (Philip Carey Corp.)*

2. Require a smooth, dry, clean substrate, free of projections that might puncture the felts. Take precautions on wood and precast decks to prevent bitumen drippage [see Chap. 3, "Structural Deck" (Design Alert No. 4)].

3. Make sure insulation is firmly attached to substrate.

4. To assure good adhesion between substrate and membrane when temperature drops below 40°F, require the following precautions:

   a. No overheating of bitumen to compensate for rapid chilling.

   b. Insulated buckets to carry the hot bitumen.

   c. Manual mopping no farther than 5 ft in front of the felt rolls and immediate unrolling of felts.

   d. Immediate application of top pour and aggregate, before the stored heat in the membrane dissipates.

   e. Below 32°F, store felts (especially coated felts) in a warm enclosure or preheat the felts before application.

5. Prohibit use of heavy mechanical roof construction equipment that may puncture the membrane or deflect the deck excessively.

6. Prohibit use of asphalt-saturated felt with coal tar pitch and vice versa.

7. Require uniform bituminous mopping, without felt touching felt, applied as follows:

   Asphalt: 20 lb per square between plies; 60-lb top coat.

   Coal tar pitch: 25 lb per square between plies; 75-lb top coat.

   Tolerance: 15 percent (*Note:* Weight tolerance of bitumen applied in extreme summer heat or winter cold may exceed +15 percent, but not −15 percent.)

8. Require a visible thermometer and thermostatic controls on all kettles, set to the following temperature limits:

Bitumen	*Minimum temp.,* °F	*Best range,* °F	*Maximum temp.,* °F	*Storage temp.,* °F
Coal-tar pitch				
ASTM D450, Type A	300	325–375	400	350
Asphalt				
ASTM D312, Type I	300	350–400	425	350
Asphalt				
ASTM D312,				
Types II, III, IV	350	400–430	475	425

Require rejection of bitumen heated above the specified maximum and reheating of bitumen too viscous for mopping. Avoid prolonged storage of bitumen in heated container. (It can lower the softening point by 10 to 20°F.)

9. Require manual brooming of all felts (to prevent air pockets and fishmouths) immediately following the mopper or felt-laying machine.

10. Prohibit phased application in which saturated felts are left exposed overnight or longer before top plies or surfacing is applied.   Place final surfacing on same day as felts.

11. Limit moisture content for surfacing aggregates as follows:
Crushed stone, gravel: 0.5 percent (by weight)
Roofing slag: 5.0 percent

12. Move dry aggregate to roof at rate of application, without stockpiling.   If unavoidable, stockpile in small mounds or rows on completed roof area, but *not on bare felts*.   Do not leave overnight.   Do not apply dusty, wet, or snow-covered aggregate.

13. If test cuts are specified, require the sampling procedure to conform with the latest ASTM- or industry-recommended practice.

14. Start roofing application at far points on deck and work toward area where materials are hoisted to roof deck (to minimize traffic over newly applied roofing).

15. On slopes of 2 in. per ft or less, apply felts perpendicular to the slope starting at the low point.

16. Backnail felts on steep asphalt roofs over $1\frac{1}{2}$ in. in slope, and on coal-tar pitch or dead-level asphalt roofs over $\frac{1}{2}$ in. in slope.

ALERTS FOR COLD-PROCESS MEMBRANES

Design

1. Do not specify the chemical-type emulsion for surfacing cold-process bitumen; specify clay-type emulsion where an emulsified bituminous surfacing is used.

2. Limit cold-process roofs to minimum $\frac{1}{2}$-in. slope.

Field

1. Do not apply cold-process emulsion surfacing at a temperature below 32°F or when there is a possibility of rain before the emulsion sets.

2. Require prior cutting and flattening of felts in cold-process membrane.

SEVEN

Flashings
and Accessories

FUNCTION AND
REQUIREMENTS

Flashings seal the joints at gravel stops, curbs, vents, parapets, walls, expansion joints, skylights, drains, built-in gutters, and other places where the built-up membrane is interrupted or terminated.

Base flashings are essentially a continuation of the built-up roofing membrane, at the upturned edges of the watertight tray. They are normally made of bitumen-impregnated, plastic, or other nonmetallic materials and applied in an operation separate from the application of the membrane itself. *Counterflashings* (or cap flashings), normally made of sheet metal, shield the exposed joints of base flashings (see Figs. 7-1 to 7-3).

Base and counterflashing combinations come in two basic types:

- Vertical terminations
- Roof edges or eaves

Flashing leaks are the most common mode of roofing failure. Whenever leaks of unknown source occur, the flashings should be

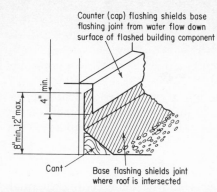

Counter (cap) flashing shields base flashing joint from water flow down surface of flashed building component

8" min.,12" max.

4" min.

Cant

Base flashing shields joint where roof is intersected

FIG. 7-1   *Base and counter-flashing.*

FIG. 7-2   *Counterflashing, erroneously omitted on left side of brick-wall control joint, shields joint in composition base flashing.*

FIG. 7-3   *Counterflashing shown above does not properly lap base flashing.*

checked first. A study of 163 roofs in southern California revealed that 68 percent of the leaks resulted from flashing failure, compared with 13 percent attributable to membrane failure. Roofing experts generally agree that faulty flashings cause from one-quarter to three-quarters of all roofing failures.

These failures stem from faulty design as well as poor field application and defective materials. Thus flashings demand as much of the designer's attention as other roofing components; detailing cannot be left to the contractor.

Flashing requires:

- Flexibility, for molding to supports and accommodating thermal, wind, and structural movement
- Compatibility with the roofing membrane and other adjoining surfaces, notably in coefficient of thermal expansion
- Resistance to slipping and sagging
- Durability, notably weather and corrosion resistance (flashings should last at least as long as the built-up membrane)

MATERIALS

Flashing materials span a broader spectrum than built-up membrane materials. They include a wide variety of metals plus the same felt and bituminous laminations used in fabricating built-up membranes and the new plastic elastomeric sheets. With the repair of leaking and deteriorating flashings so common and costly, first-cost economy should be a secondary consideration in material selection and design; quality and durability are the prime considerations in flashing design.

The use of nonmetallic materials for base flashings and metals for counterflashings exploits the best qualities of each. Because it has a similar coefficient of thermal expansion, plied felt or fabric-based flashing works best with the built-up membrane as base flashing. And, due to their good weather and corrosion resistance, metal cap flashings (or counterflashings) provide superior surface protection. As long as the base and counterflashings are designed to allow relative movement, they can function together despite their different thermal expansion coefficients (see Table 7-1).

Because of their rigidity, their incompatible coefficients of thermal expansion, and the difficulty of connecting them to the roof membrane, metal flashings are generally unsuitable as base flashings (see Fig. 7-4). A notable exception to this rule is the base flashing for a roof drain where the need for stability and a good bolted connection to the metal drain

table 7-1   THERMAL EXPANSION OF COUNTERFLASHING METALS

Metal	Thermal expansion coefficient $\times 10^{-6}$ in./(in.)(°F)	Expansion of 10-ft length for 100°F temperature rise (in 64ths in.)
Galvanized steel..............	6.7	5
Monel.....................	7.8	6
Copper....................	9.4	8
Stainless steel (300 series).....	9.6	8
Aluminum..................	12.9	10
Lead.......................	15.0	12
Zinc, rolled.................	17.4	13

frame favors lead, copper, or other sheet metal. Wherever possible, skylights, ventilators, and similar roof-penetrating elements, which are sometimes detailed for metal base flashing, should be set on curbs to which membrane base flashings can be attached.

The designer's choice of flashing materials and details is often limited under a manufacturer's roofing bond or flashing bond. Under these bonds, the manufacturer normally requires use of his materials.

Base Flashings

Base flashings are generally composition flashings—a generic term denoting any nonmetallic, bituminous flashing, including fabric or felt, applied with hot-mopped asphalt or cold-process bituminous plastic flashing cement.

Fabrics, woven from fiberglass or cotton, tend to offer greater strength per ply than felted flashings. Close-woven fabrics are more puncture-

FIG. 7-4 *Metal base flashing shown is bad design, requiring a difficult connection with the built-up membrane. Composition base flashing with a cant strip would have averted a vulnerable continuous joint between membrane and flashing.*

resistant, more flexible, and lighter, and thus more easily molded to flashed surfaces than felted flashings.

Felted flashings are generally more durable than fabric flashings. Some laminated composition flashings exploit fabric and felt for their complementary qualities; reinforced asbestos felt on the exposed surface gives good durability, while fabric backing provides strength.

Felted flashings are generally made of the same number of plies as the built-up membrane. Fabric impregnated with a bituminous material or mastic and bonded with plastic cement sometimes has fewer plies than the membrane.

Interply coatings of flashings may be either hot-mopped steep asphalt or cold-applied plastic cement, troweled or brushed on the flashing material.

Chief advantages of hot-mopping are speed, convenience, and economy. One disadvantage of hot-mopped asphalt is the chance of mistakenly applying bitumen of lower softening point, resulting in sagging of the base flashing.

Unlike hot-mopped asphalt, cold-applied plastic cement can serve both as an interply waterproofing and adhesive and as a surfacing material. (Asbestos fibers give plastic cement its stability.) Fabric and organic felt flashings must be surfaced with plastic cement or clay-type asphalt emulsion. Mineral-surfaced roll roofing, often placed over hot-mopped flashing felt, requires no such surfacing.

The top edge of composition base flashing is sometimes counterflashed with similar bituminous materials. When such counterflashing covers a large area, it is known as a waterproofing wall system. When it covers only a narrow area, it is a "stripping" or "seal."

Accessories

Accessories provide a protective termination for the attached built-up composition roofing and base flashing. They must be weathertight, water-shedding, and wind-resistant. In the broad sense, accessories include (1) edgings and (2) counterflashings.

Edgings   Edgings (normally made of metal) include gravel stops, coping covers, fascia strips, rake strips, combination fascia-gutter-gravel-stop assemblies, and similar roof-terminating devices. Metal edgings are generally tied into the built-up roofing membrane before the final surfacing is applied by placing the flange on the membrane surface and sealing the metal lap with composition stripping. An effective seal consists of a strip of two or more plies of fabric or felt set into and troweled with flashing cement.

Combinations of fabrics (cotton and glass) and felts (asbestos or organic) are chosen to provide varying degrees of elasticity, strength, and durability in five- and seven-course seals.

The flange of a metal accessory tied into a built-up roofing membrane should be at least 4 in. wide. It should be primed and set into place on the membrane surface in a bed of flashing cement, fastened to a nailing strip with nails or screws spaced not more than 4 in. apart. Except where the edge detail forms a gutter, the nailing strip surface should be raised above the anticipated waterline.

Counterflashings   Counterflashings shield the exposed joints of base flashings and shed water from vertical surfaces onto the roof. Because of their exposed locations, counterflashings must be rigid and durable; thus metal generally proves the best counterflashing material. Some nonmetallic counterflashings are made of asbestos felts, and are water-proofed with highly fibrated asbestos flashing cement.

Vinyl, neoprene, butyl rubber, nonplasticized chlorinated polyethylene reinforced with polyester mesh, and other new elastomeric flashings are applied in one ply and readily molded to building contours. Adequate history of their field performance is, however, not yet available. Among troubles already encountered with some of these materials are deterioration in sunlight and gradual loss of plasticizer. Cold-weather shrinkage, embrittlement, and lack of flexibility can result in cracking.

Costly failures have resulted from specifying unsuitable flashing materials. Designers should insist on technical data from laboratory and field tests before accepting these new materials.

PRINCIPLES OF FLASHING DESIGN

Good flashing design is based on several principles:

- Allowing for differential movement between base and cap flashings anchored to different parts of the structure.
- Locating flashed joints above the highest water level on the roof.
- Contouring flashed surfaces to avoid sharp bends in bituminous base flashings.
- Connecting flashings solidly to supports.

Differential Movement

A basic principle of flashing design is to provide for differential movement among the different parts of the building. Anchor base flashing to the

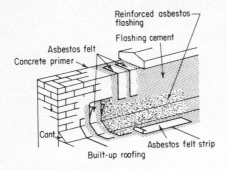

FIG. 7-5   *The use of felted counter-flashing (shown in drawing), without provision for differential movement between roof deck and wall, is satisfactory only if the roof framing bears on the wall or is otherwise anchored to it, thus preventing separation.*

structural roof deck, free of the walls or other intersecting elements, and anchor the counterflashing to the wall, column, pipe, or other flashed element.

This need to accommodate relative movement between structurally independent building elements is often overlooked—notably in flashing details that incorrectly connect base flashing direct to the walls (see Figs. 7-5 and 7-6). Since it is difficult to calculate the magnitude and precise direction of these relative movements, the designer should provide conservative allowances in his details.

Differential movement must also be accommodated at gravel stops. Membranes stripped in with metal gravel stops often split above joints in metal (see Fig. 7-7). In extreme climates, notably the Midwest, the annual roof surface temperature may range 180°F (from −20 to 160°F). In a 10-ft aluminum fascia strip, this temperature differential produces a total lateral deformation of $\frac{1}{4}$ in. or more. Under annual cycling, this movement can split a membrane bonded tight to the metal fascia.

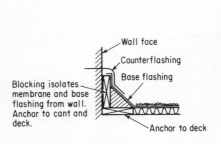

FIG. 7-6   *A good wall-flashing detail interposes wood blocking, anchored to the deck, to allow differential movement between the deck (base flashing) and wall (counterflashing).*

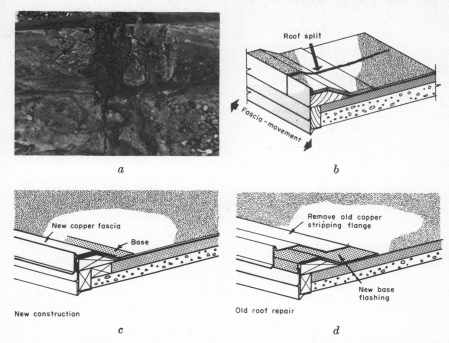

FIG. 7-7 *Bonding the roof membrane direct to the metal flange of a gravel stop can split the membrane when the metal contracts at low temperatures (see a and b). Isolating the membrane from the metal fascia by anchoring it to stable, treated lumber can prevent local edge splitting (see c and d).* (Tamko Asphalt Products.)

Isolating the membrane from the metal fascia by using a stable cant strip of treated lumber can prevent splitting.

Flashing Elevation

Another occasionally violated principle of good flashing design is locating joints between roofing and flashing above the general roof water level. To reduce the risk of leakage at these vulnerable joints, the designer should slope the roof upward toward peripheral gravel stops. He should coordinate the drainage design with the flashing layout—to place vent stacks, wall intersections, and other flashing locations at high points. At the least, he should keep flashings out of low areas where ponding might occur.

Base flashing should extend at least 8 in. above the highest anticipated waterline. Metal counterflashing should lap the base flashing at least 4 in.

Flashing Contours

A third basic rule of flashing design is not to bend bituminous base flashing to fit right-angle corners. To reduce the risk of cracking use cant strips of 45-degree maximum bend.

Flashings should be supported continuously—not left to bridge openings. Where such spanning by cap flashing is unavoidable, e.g., at expansion joints and parapet walls, detail flashing for maximum strength.

Flashing Connections

Anchorage for flashing is generally the same as for the built-up membrane and subject to the same limitations. The importance of secure nailing for flashings is even greater than for the membrane because flashings are more vulnerable to wind uplift and other physical damage. To prevent backout, specify annular ring or screw-type nails. If driven through metal flashing, nails should be compatible with the flashing material to avoid galvanic corrosion or staining.

L nailers must also be securely anchored to the deck (see Fig. 7-8).

SPECIFIC CONDITIONS

Flashings can be classified by function as:

1. Wall, or other vertical intersections
2. Edge details, e.g., gravel stops, eave flashings
3. Expansion joints
4. Roof penetration connections—for vents, skylights, roof drains, scuttles, pitch pockets, stub columns, etc.
5. Water conductors, e.g., built-in gutters, valleys, scuppers

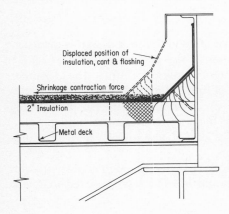

FIG. 7-8 *Insecure flashing and nailing strips, plus sliding insulation, can be dragged several inches by a contracting built-up roofing membrane. (National Bureau of Standards Monograph No. 89.)*

Each type shares the general problems previously discussed, but also has its own peculiar problems.

Wall Flashings

Differential movement between wall and membrane is almost inevitable. To isolate the vertical surface of the base flashing from the wall, place an upright blocking board behind the cant strip as a back surface for the base flashing. Like the cant strip, this blocking should be anchored to the structural deck. Counterflashing shields the open joint of the base flashing (see Fig. 7-6).

To prevent water from filtering through the core of the masonry wall, use through-the-wall flashing rather than a reglet, wherever possible. If a reglet cannot be avoided—e.g., in a concrete wall—use a prefabricated cap flashing system. Avoid use of a prefabricated cant and reglet block system into which the membrane flashing is inserted.

Caulking of reglets, to seal the joint where the cap flashing is inserted, is critical. It requires a good sealant—not plastic cement, oil-based caulking, or lead wood fillers.

Because of the difficulty in flashing them, parapet walls are best eliminated unless there are compelling architectural reasons for their use.

Edge Details

Gravel stops perform triple duty—as edge termination for the membrane, as rain shield, and as top surface of a fascia strip; thus they require a rigid metal or plastic material. End laps, expansion joints, and connection to the roofing membrane must all be watertight and flexible to allow for thermal movement and for differential movement of dissimilar materials.

FIG. 7-9 *Plastic cement surfacing displays random shrinkage cracking plus longitudinal crack (finger) where fabric base flashing terminates. Counterflashing should shield base-flashing joint. Random shrinkage cracking can result from excessively thick flashing cement.*

Gravel stops should be raised above the general roof level to prevent water from penetrating the roof edge. The base flashing should be set in plastic cement and stripped to the roofing membrane with two layers of bituminous-impregnated woven fabric or felt covered with plastic cement.

Edge details are critical in resisting wind uplift. Most wind blowoffs start with wind penetrating the roof-edge detail, often bending and twisting the metal fascia strip and then rolling back the membrane. As safeguards, the designer should specify minimum 18-gage metal for the fascia and nailing at 3- or 4-in. maximum spacing. For metal deck roofs, edge flashing details should prevent wind access into the fluted sections under the membrane on the upper deck surface.

Expansion Joints

Expansion joints, generally spaced from about 150 to 300 ft, relieve stresses that would otherwise accompany thermal expansion and contraction. They should be provided at changes of span direction in the structural deck or framing, or at changes in deck material.

To accommodate free structural movement at an expansion joint, the flashing seal must permit movement across the joint and shear movement parallel to the joint. The design must also permit longitudinal movement in metal counterflashings to prevent compressive buckling when a heated flashing is restrained from expanding (see Fig. 7-10).

Expansion-joint details featuring a fold in the roofing membrane may behave like a roof wrinkle, with bitumen flowing down the slope of the fold. In such detail, fatigue bending often cracks the weakened membrane. Here the new elastomeric materials are superior.

Roof Penetrations

The best flashing detail at roof penetrations follows the general rule of attaching the base flashing to the structural deck and counter flashing to the penetrating element. Avoid pitch pockets. They pose a constant open threat of leaks and require frequent maintenance checks and repair of bitumen embrittled by sunlight and cold weather. Wherever practicable, build pedestals for roof-penetrating elements and flash them.

To avoid the cost of providing separate support for base flashing at small pipes, vents, and similar roof penetrations, designers often omit the separate support for base flashings (see Fig. 7-12). Instead, the flanges of metal or plastic sleeves are attached direct to the roofing membrane, and the flanged sheet metal support carries the base flashing around it. Counterflashing attached to the pipe shields the base flashing. As a secondary defense against leakage, the flanged sheet metal support, or "pan," is filled with soft bituminous mastic.

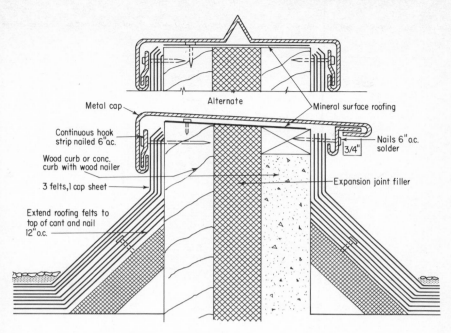

FIG. 7-10 *Typical roof expansion-joint detail.*

FIG. 7-11 *Fabricating a sheet-metal box, detailed for lateral entry of pipes or conduit (left), is far better practice than allowing vertical entry into a pitch pocket (right).*

Such details are normally satisfactory, but, like pitch pockets, require more frequent inspection and maintenance than the basic flashing detail with deck-supported base flashing. Support on membrane and insulation is less stable than support on the structural deck and thus more vulnerable to leakage resulting from joint separation.

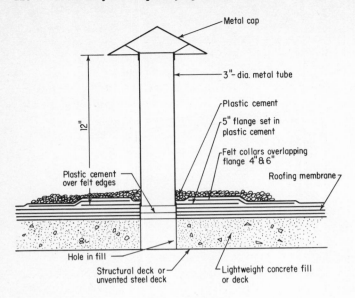

Moisture release vent

FIG. 7-12   *Small vent detail.*

Since clogged drains can impound water for long intervals, good flashing is especially important at drains. Metal sheet provides the best connection to drain frames; it also provides the stability needed to maintain a watertight joint.

Drains are also threatened with differential vertical movement between drain pipe and roof. A structural roof deck supported on shrinking 2- × 12-in. wood joists may lower the deck elevation ½ in., breaking or distorting the drain flashing or its connection. An expansion joint or a 45-degree offset in the drainpipe can accommodate most differential vertical movement.

Water Conductors

Built-in gutters are subject to surface abrasion. They are also difficult to connect to the membrane of the drained roof surface. For relatively flat gutters, which may retain water for long intervals, leakage in sliding expansion joints also poses a problem.

Metal offers superior durability but, largely because of its high thermal coefficient of expansion, is difficult to connect to the bituminous membrane. Gutters made of bituminous materials readily accommodate thermal movement, but they deteriorate so rapidly under wetting and drying cycles and the abrasive flow of water that they are limited to very short valleys.

ALERTS

General

1. Check with roofing manufacturer for approval of all flashing details and materials, including primers, bitumens, fabrics, coated base sheets, surfaced cap sheets, and bituminous mastic for use with flashing details.

2. Supplement specifications with large-scale flashing details on drawings.

Design

1. Limit number of penetrations through the roof. Group roof-mounted equipment requiring flashing with pads, curbs, etc., into a smaller number of larger areas to reduce total quantity of flashing. Provide minimum leg height for rooftop equipment supports per Table 7-2.

table 7-2   MINIMUM SUPPORT
HEIGHT FOR ROOFTOP
EQUIPMENT

Equipment width, in.	*Minimum support height, in.*
Up to 24	12
25–36	18
37–48	24
49–60	30
60 and wider	36

* When two or more pieces of equipment are installed side by side—less than 36 in. apart—they shall be considered as one piece of equipment and the leg height so determined.

2. If feasible, eliminate parapet walls.

3. Wherever parapets or other walls intersect the roof, provide through-the-wall flashing in preference to reglets.

4. In general, avoid metal base flashing; use composition flashing.

5. Avoid right-angle bends in base flashing; use cant strips to provide an approximate 45-degree angle.

6. Never anchor base flashing directly to walls, structural members, pipes, or other building elements independent of the roof assembly. Always provide for differential movement.

7. Where metal cap flashing overlaps base flashing, detail joint for differential movement. Locate these joints above highest water level expected on roof.

8. Limit vertical extension of base flashing to a nominal 8 in. minimum, 12 in. maximum, above membrane grade line. Protect back of parapet walls and similar surfaces with metal counterflashing.

9. Preservative treatment, compatible with bitumen, of wood cant strip, nailers, and blocking used with roof assembly may help in many areas.

10. Avoid pitch pockets for columns and other structural members penetrating the roof. Build a pedestal and cap flash it.

11. Specify minimum 18-gage sheet metal for gravel stops (to resist wind bending). (Lighter-gage material may prove satisfactory if the detail provides a lock strip at the lower edge of the fascia or if fasteners can penetrate the fascia.)

12. Specify an expansion joint between drain and leader pipes, or offset the leader pipe 45 degrees. (When these units are rigid, movement can break flashing connections and permit water entry into membrane.)

Field

1. Establish a maintenance program with annual inspection followed by resurfacing of deteriorating surfaces, remopping of opened joints, renewal of all joint and reglet caulking, and repair of rips, tears, and other defects.

2. Install flashing, including counterflashing, as roof application progresses. If delay is unavoidable, trowel the top of the flashing with flashing cement to close the joint and prevent water from entering behind the flashing until the counterflashing is in place.

EIGHT

Fire Resistance

Fire is the most costly and hazardous destructive force that can attack a built-up roof system. Wind-uplift failures, though still less costly, have been increasing at a faster rate than fire losses. Other insured roof losses, e.g., earthquake or hail damage, are generally less significant than either fire or wind.

The roof designer's chief concern in designing against fire and wind damage is to satisfy the building code. His second concern, almost equally important, is satisfying the insurance company's conditions for qualifying the building for extended coverage. Use of a combustible instead of a fire-resistive roof assembly could raise the owner's overall building cost by making a costly sprinkler system mandatory. The cost of sprinklers will outweigh any conceivable saving in substituting a cheap combustible roof for a more expensive fire-resistive roof.

The most important standards-setting organizations for fire and wind-uplift resistance are the Underwriters' Laboratories, Inc. (UL), of Northbrook, Illinois, an independent, nonprofit testing organization, and the Factory Mutual Engineering Corporation and Association (FM) of Norwood, Massachusetts. These organizations classify roof assemblies for fire and wind-uplift resistance for the nation's insurance companies.

Both UL and FM maintain laboratories for testing and listing manufacturers' building products satisfying their standards. Insurance companies follow their approved construction listings. UL and FM standards also serve as the basis for building code requirements for fire and wind-uplift resistance of built-up roof assemblies.

The following discussion first describes the testing agencies' programs. An understanding of basic testing philosophy and techniques is essential to understanding building code provisions.

HISTORICAL NOTE

The fire that destroyed a huge General Motors plant at Livonia, Michigan, in 1953 had a tremendous impact on fire-resistive standards for steel roof decks. Because they represent a larger annual construction volume than all other deck materials combined, steel decks are especially important. Representing the nation's greatest fire insurance loss (until the 1966 McCormick Place fire), the Livonia fire gave the industry a stark lesson on the fire-feeding hazard of hot-mopped bitumen applied direct to a steel deck. It also influenced wind-uplift design, inspiring a search for cold-applied adhesives and mechanical fasteners to replace the hazardous bituminous moppings previously applied direct to steel decks with combustible insulation.

FIRE TESTS AND STANDARDS

Following the Livonia fire, tests conducted in a 20- $\times$ 100-ft standard test building by FM and UL confirmed the hazards of placing large quantities of bitumen direct on a steel deck with combustible wood-fiber insulation. The bitumen and wood fibers spread test fires through the 100-ft-long building within 10 to 12 min, regardless of variations in the number of moppings (from one to three) between the deck and the insulation. The full-scale building tests conclusively demonstrated that no hot bituminous mopping of sufficient thickness to bond the insulation could be part of a fire-resistive steel deck roof system with wood fiber or similar insulation.

Noncombustible insulation—foamed glass, glass fiber, lightweight insulating concrete—proved satisfactory with limited amounts of hot-mopped bitumen applied direct to the steel deck.

Flame-spread Test

To satisfy building code and insurance requirements, a roof system must always resist internal fire (fire within the building). (Some insurance companies ignore the less serious external fire threat.) The chief safeguard required against internal fire is limitation of flame spread along the underside of the roof assembly.

Resistance to flame spread is tested by an adaptation of the "Test Method for Fire Hazard Classification of Building Materials" (UL723 and ASTM E84). The test roof assembly forms the top of a test tunnel in which twin gas burners deliver flames against its underside. The gas supply and other variables are adjusted until the furnace produces a flame-spread rate of 19½ ft in 5½ min on select-grade red-oak flooring.

As a result of the Livonia fire, the steel deck became standard roof-deck construction for the classification of roof assemblies' resistance to flame spread, as measured by the flame-spread test. As its standard for rating roof-deck constructions, UL "Building Materials List" chose a lightgage steel roof deck, insulated with 1-in.-thick plain vegetable-fiber board. The insulation is mechanically anchored and covered with a four-ply, gravel-surfaced built-up roofing membrane. To qualify for listing as "acceptable" by UL, a roof-deck assembly must not spread the flame of a modified UL723 flame-spread test farther than the standard steel deck assembly described above.

In the basic 20- × 100-ft test building, a roof assembly qualifying for a UL fire rating of "acceptable" must not spread flame more than 60 ft from the fire end of the test structure on the deck underside during the 30-min test. For the alternative, more economical, tunnel test, an acceptable roof assembly must not spread flame on the underside more than 10 ft during the first 10 min, and 14 ft during the next 20 min (see (Fig. 8-1).

FM Fire Classification

Factory Mutual's classification of resistance to interior fire divides roof assemblies (deck, vapor-barrier, insulation) into two basic categories: sprinklered and unsprinklered. (These classes refer to the roof construction for nonhazardous occupancies, since some interiors containing combustible materials require sprinklers *regardless of the fire resistance of the roof system.*)

Roof assemblies *not* requiring sprinklers include:

- Noncombustible decks—concrete, gypsum, asbestos cement, and preformed structural mineralized wood fiber
- Wood decks treated with fire-retardant, inorganic salts limiting flame spread to 25 or less
- Class I steel decks

Roof assemblies requiring sprinklers include:

- Combustible decks (untreated wood)
- Class II steel decks

FIG. 8-1   *To qualify as a Class I steel-deck roof system, a specimen roof assembly (roof of test tunnel) must not spread flame from twin gas burners more than 10 ft in the first 10 min, or 14 ft during the next 20 min.   (Underwriters' Laboratories, Inc.)*

The distinction between Class I and Class II steel decks depends upon (1) the method of bonding preformed insulation to the deck surface, and (2) the fire resistance of the insulation itself.

The use of hot-mopped bitumen on a steel deck surface to form a vapor seal or to bond combustible, organic insulation to a built-up membrane disqualifies a steel deck for Class I rating.   One exception to this rule occurs when a deck soffit is sprayed with a noncombustible insulation. Otherwise, only FM-approved adhesives or approved mechanical fasteners can qualify a steel deck for a Class I rating.   (For approved manufacturers' adhesives and fasteners, see FM "Loss Prevention Data," FM "Construction 1-28S," latest edition, and UL "Building Materials List," under Roof Deck Constructions.)

The application of foamed plastic insulation (polystyrene or urethane) direct to its top surface also disqualifies a steel deck roof assembly for a Class I rating (though a breakthrough may be imminent).   Subjected to heat from an interior fire, foamed plastic disintegrates and exposes the bitumen of the built-up membrane to heat conducted by the steel deck.   It thus creates the same fire-feeding hazard as a bituminous mopping.   A UL-listed foamed plastic insulation requires a lower layer of another insulating material sandwiched between the steel deck and the plastic foam.   (See UL "Building Materials List," Roof Deck Construction No. 9.)

Time Fire Rating

Under many building codes a roof assembly must have a fire rating established by its performance in a standard ASTM E119 furnace test. The "National Building Code," recommended by the American Insurance Association, requires a 2-hr fire rating for roofs in fire-resistive Type A construction. To qualify for this fire rating, a tested roof assembly (or any other tested system or component) must endure a test fire of progressively rising temperature—from 1000°F at 5 min to 1700°F at 1 hr and 1850°F at 2 hr (see Fig. 8-2). This standard test is promulgated by three major organizations: Underwriters Laboratories (UL263); American Society of Testing and Materials (ASTM E119); and the National Fire Protection Association (NFPA 251).

The fire-endurance test measures roof-assembly performance in carrying loads and in confining fire. To qualify for a given fire rating, the tested assembly must (1) sustain the applied design load; (2) permit no passage of flame or gases hot enough to ignite cotton waste; and (3) limit the average temperature rise of the unexposed face to a maximum 250°F above its initial temperature or a 325°F rise at any one point.

Fire-resistance ratings of various roof-ceiling assemblies appear in UL "Building Materials List" under Fire Resistance Classifications, Roof and Ceiling Constructions.

Roof Covering Ratings

To measure their resistance to external fire, roof coverings are tested and rated. The main fire-resistive function of the built-up membrane is to reduce the risk of fire spreading from neighboring buildings. The roof

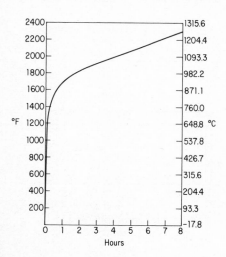

FIG. 8-2 *Standard time-temperature curve (ASTM E-119) shows rising temperatures that construction assemblies must endure to qualify for fire-resistance ratings (in hours).*

covering should not spread flame rapidly, produce flying brands endangering adjacent buildings, or permit ignition of combustible roof decks.

According to the industry standard for rating roof coverings, "Test Methods for Fire Resistance of Roof Covering Materials," UL790, classified roof coverings "are not readily flammable, do not slip from position, and possess no flying brand hazard." Their performance is rated for different fire intensities:

- Class A roof coverings "are effective against *severe* fire exposure."
- Class B roof coverings "are effective against *moderate* fire exposure."
- Class C roof coverings "are effective against *light* fire exposure."

Still another distinction among roof-deck coverings is the degree of protection they give the roof deck. Under "severe fire exposure," a Class A roof covering affords *"a fairly high* degree of fire protection"; under moderate fire exposure, a Class B roof covering affords *"a moderate degree* of protection"; and under light fire exposure, a Class C deck affords *"a measurable degree* of fire protection."

To determine these classifications, UL conducts three tests:

- Flame exposure
- Flame spread
- Burning brand

The testing apparatus consists of a 3-ft-long gas burner placed between a large air duct and the roof-covering specimen, which is mounted on a standard timber deck and set at the maximum slope for which the tested covering is recommended by the manufacturer (see Fig. 8-3). In all three tests an air current of 12 mph blows across the test specimen.

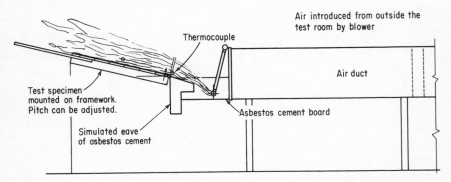

FIG. 8-3   *Testing apparatus for qualifying roof coverings through* flame exposure, flame spread, *and* burning brand *tests.   (Underwriters' Laboratories, Inc.)*

The *flame-exposure test* subjects the specimens to varying cycles of flame exposure at 1400°F for Class A and B tests and 1300°F for Class C.

The *flame-spread* test subjects a 13-ft-long specimen to a similar gas flame for either 10 min (Class A and B), 4 min (Class C), or until the flame recedes from the point of maximum spread.

The *burning-brand test* specifies different sizes of test brands simulating the firebrands hurled by wind or by uprushing gases of an actual fire onto the roofs of nearby buildings. The test-brand sizes vary from a 12-in.-square lumber lattice for Class A down to a 1½-in.-square block for Class C (see Fig. 8-4). In these tests, the brands are ignited in a gas flame at temperatures over 1600°F for varying periods of time, depending on the classification sought. They are placed on the tested roof assembly and allowed to burn until consumed.

To qualify for classification, the roof covering must withstand the three tests (*flame exposure, flame spread,* and *burning brand*) (1) without any portion of the roof covering material blowing or falling off in glowing brands; (2) without exposing the roof deck by breaking, sliding, cracking, or warping; and (3) without permitting ignition or collapse of any portion of the roof deck. (Other conditions of varying severity are established in "Test Methods for Fire Resistance of Roof Covering Materials," UL790 or ASTM E108.)

The classifications of built-up roofing membranes are published in the UL "Building Materials List," under Built-up Roof Covering Materials. Most Class A and Class B built-up roofing membranes are aggregate-surfaced, but some smooth-surfaced asbestos felt membranes and mineral-surfaced cap sheets also qualify. Felts for use in these UL-listed built-up membranes carry the UL label.

BUILDING CODE PROVISIONS

Building code provisions for roofs, as well as other building components, almost universally follow FM or UL standards. All four model codes, the basis for most of the nation's local codes, classify buildings by degree of fire resistance and occupancy.

Under the BOCA "Basic Building Code," promulgated by the Building Officials Conference of America (BOCA), the degree of fire resistance is defined by four basic types of construction:

- Type 1: fireproof
- Type 2: noncombustible
- Type 3: exterior masonry walls
- Type 4: frame

table 8-1 GENERAL HEIGHT AND AREA LIMITS OF BUILDINGS FACING ON ONE STREET OR PUBLIC SPACE NOT LESS THAN 30 FT WIDE

(From BOCA "Basic Building Code" 1965 ed.)

(Areas in square feet; heights in number of stories and feet)

| | Use group | Type 1 Fireproof | | Type 2 Noncombustible | | | Type 3 Exterior masonry walls | | | Type 4 Frame | |
| | | | | Protected | | Unprotected | Heavy timber (mill) | Ordinary joisted Protected | Ordinary joisted Unprotected | Protected | Unprotected |
		1A	1B	2A	2B	2C	3A	3B	3C	4A	4B
A	High hazard — Notes h and m*	5St.65' 16,800	3St.40' 14,400	3St.40' 11,400	2St.30' 7,500	1St.20' 4,800	2St.30' 7,200	2St.30' 6,600	1St.20' 4,800	1St.20' 5,100	NP‡
B-1	Storage (moderate) — Notes b, d, e, f, i, and j*	U†	U	5St.65' 19,950	4St.50' 13,125	2St.30' 8,400	4St.50' 12,600	3St.40' 11,500	2St.30' 8,400	2St.30' 8,925	1St.20' 4,200
B-2	Storage (low) — Notes b, d, e, and f*	U	U	7St.85' 34,200	5St.65' 22,500	3St.40' 14,400	5St.65' 21,600	4St.50' 19,800	3St.40' 14,400	3St.40' 15,300	2St.30' 7,200
C	Mercantile — Notes b, d, e, and f*	U	U	6St.75' 22,800	4St.50' 15,000	3St.30' 9,600	4St.50' 14,400	4St.50' 13,200	2St.30' 9,600	3St.30' 10,200	1St.20' 4,800
D	Industrial — Notes b, d, e, and f*	U	U	6St.75' 22,800	4St.50' 15,000	3St.30' 9,600	4St.50' 14,400	4St.40' 13,200	2St.30' 9,600	2St.30' 10,200	1St.20' 4,800
E	Business — Notes b, d, e, and f*	U	U	7St.85' 34,200	5St.65' 22,500	3St.40' 14,400	5St.65' 21,600	4St.50' 19,800	3St.40' 14,400	3St.40' 15,300	2St.30' 7,200
F-1-A	Assembly theaters: with stage and scenery	U	6St.75' 14,400	4St.50' 11,400	2St.30' 7,500	1St.20' 4,800	2St.30' 7,200	2St.30' 6,600	1St.20' 4,800	1St.20' 5,100	N.P.
F-1-B	Assembly Theaters: without stage (motion picture theaters)	U	U	5St.65' 19,950	3St.40' 13,125	2St.30' 8,400	3St.40' 12,600	3St.40' 11,550	2St.30' 8,400	1St.20' 58,92	1St.20' 4,200

Type of construction

Use	Description										
F-2	Assembly: night clubs and similar uses	U	4St.50' 7,200	3St.40' 5,700	2St.30' 3,750	1St.20' 2,400	2St.30' 3,600	2St.30' 3,300	1St.20' 2,400	1St.20' 2,550	1St.20' 1,200
F-3	Assembly: lecture halls, recreation centers, terminals, restaurants other than night clubs Note e*	U	U	5St.65' 19,950	3St.40' 13,125	2St.30' 8,400	3St.40' 12,600	3St.40' 11,550	2St.30' 8,400	1St.20' 8,925	1St.20' 4,200
F-4	Assembly: churches, schools Note n and o*	U	U	5St.65' 34,200	3St.40' 22,500	2St.30' 14,400	3St.40' 21,600 Note g	3St.40' 19,800	2St.20' 14,400	1St.20' 15,300 Note g	1St.20' 7,200 Note g
H-1	Institutional: restrained	U	6St.75' 18,000	4St.50' 14,250	2St.30' 9,375	1St.20' 6,000	2St.30' 9,000	2St.30' 8,250	1St.20' 6,000	1St.20' 6,375	N.P.
H-2	Institutional: incapacitated	U	8St.90' 21,600	4St.50' 17,100	2St.30' 11,250	1St.20' 7,200	2St.30' 10,800	2St.30' 9,900	1St.20' 7,200	1St.20' 7,650	N.P.
L-1	Residential: hotels	U	U	9St.100' 22,800	4St.50' 15,000	3St.40' 9,600	4St.50' 14,400	4St.50' 13,200	3St.40' 9,600	3St.40' 10,200	2½St.35' 4,800
L-2	Residential: multifamily	U	U	9St.100' 22,800	4St.50' 15,000 Note k	3St.40' 9,600	4St.50' 14,400	4St.50' 13,200 Note k	3St.40' 9,600	3St.40' 10,200	2½St.35' 4,800
L-3	Residential: 1 and 2 family	U	U	4St.50' 22,800	4St.50' 15,000	3St.40' 9,600	4St.50' 14,400	4St.50' 13,200	3St.40' 9,600	St.40' 10,200	2½St.35' 4,800
M	Miscellaneous and temporary	U									

* See footnotes a through o to BOCA "Basic Building Code," Table 6, for allowable modifications to tabulated values.

† U = unlimited.

‡ NP = not permitted.

Each of these types has subclasses, e.g., Type 1, Fireproof Construction, 1A and 1B (see Table 8-1), and each subclass carries its own required fire rating for the roof assembly, e.g., $\frac{3}{4}$ hr for Type 2A, noncombustible, protected construction.

To determine what type of construction is required for his building under the BOCA building code, the designer consults Table 8-1. For any given use, the larger the building, the more stringent the fire requirements. Conversely, the more fire-resistive the construction, the larger the permitted building.

For example, any residential hotel more than nine stories high or 22,800 sq ft in area must have at least Type 1B roof assembly ($1\frac{1}{2}$-hr fire rating) when the top floor ceiling dimension is 15 ft or less (see Table 8-2).

The "National Building Code" similarly establishes roof-construction requirements—"fire resistive," "ordinary," "wood frame," etc. (see Table 8-3). These classes similarly range from a maximum required 2-hr fire rating down to zero fire-rating requirement for "wood frame" (see Table 8-4).

In accordance with the lesson of the Livonia fire, Section 705.3 of the "National Building Code" bans the use of "a metal roof deck with any material applied directly to its upper surface which represents the hazard of propagation of fire on the underside of the metal roof deck" for roofs of unsprinklered buildings that exceed 9,000 sq ft in area.

FIG. 8-4   *Test brands burned to establish UL Class A, B, or C roof-covering rating range from a 12-in.-square saw-cut block for Class C.* (*Underwriters' Laboratories, Inc.*)

table 8-2 FIRE-RESISTANCE RATINGS (HOURS) OF ROOF ASSEMBLIES AND FRAMING

(From BOCA "Basic Building Code," 1965 ed.)

| | Type 1 Fireproof | | Type 2 Noncombustible | | Note b* | Type 3 Exterior Masonry walls | Ordinary | | Type 4 Frame | |
| | | | Protected | | Unprotected | Heavy timber (mill) | Protected | Unprotected | Protected | Unprotected |
Structural element	1A	1B	2A	2B	2C	3A	3B	3C	4A	4B
11 Roof construction including beams—15 ft or less in height Note i*	2	1½	¾ i*	¾ i*	0 i*	See Sec. 217	¾	0	½	0
12 Roof trusses and framing including arches and roof deck Note f* — More than 15 ft but less than 20 ft in height to lowest member	¾ i*	¾ i*	¾ i*	0 i*	0 i*	See Sec. 217	0	0	½	0
20 ft or more in height to lowest member	0 i*	0 i*	0 i*	0 i*	0 i*	See Sec. 217	0	0	0	0

* See footnotes b, f, i, and j; to BOCA "Basic Building Code" Table 5, Section 217, for modification of tabulated values.

table 8-3   HEIGHT AND AREA LIMITS OF BUILDINGS*
(*"National Building Code"*)

Type of construction	Maximum height, ft	Maximum area, sq ft	
		One-story	Multistory
Fire-resistive, Type A...............	No limit	No limit	No limit
Fire-resistive, Type B...............	85	No limit	No limit
Protected noncombustible...........	75	18,000	12,000
Heavy timber.....................	65	12,000	8,000
Ordinary.........................	45	9,000	6,000
Unprotected noncombustible........	35	9,000	6,000
Wood frame......................	35	6,000	4,000

* For exceptions to above tabulations, see "National Building Code" Article III, Section 400.   For sprinkler requirements, see Article 812.

The "National Building Code" rates roof coverings in conformance with the UL classification: Class A, "effective against *severe* fire exposure"; Class B, "effective against *moderate* fire exposure"; and Class C, "effective against *light* fire exposure."

The code requires Class A or B roof covering on every building except the following:

- Dwellings
- Wood-frame buildings
- Buildings located outside a fire district, and qualified, on the basis of height and area limits, for wood-frame construction.

Any building exempted from the requirement for Class A or B roof covering, must, however, have at least a Class C covering.   (Under the "National Building Code," there is no distinction between the requirements for a Class A or Class B roof covering.)

FIRE ALERTS

1. Before specifying a roof assembly, consult with the local building official and insurance company representative to make sure that your design satisfies local fire requirements.

2. Beware of substituting even a single component within a fire-rated roof-ceiling assembly.   Such changes require restudy of the new roof assembly thus created.   A fire rating applies to an entire roof-ceiling system, not to its individual components.

table 8-4 ROOF CONSTRUCTION MATERIALS
AND FIRE-RESISTANCE RATINGS
(*"National Building Code,"* 1967 ed.)

Type of construction	Required roof construction
Fire-resistive, Type A	Approved noncombustible material; 2-hr fire rating
Fire-resistive, Type B	Approved noncombustible material; 1½-hr fire rating
Protected noncombustible	Approved noncombustible material; 1-hr fire rating
Unprotected noncombustible	(1) Approved noncombustible material; (2) unsprinklered buildings exceeding 9,000 sq ft in area shall not have a metal roof deck with any material, applied directly to its upper surface, that presents the hazard of fire propagation on the deck underside.
Heavy timber	(1) Matched planks of minimum 2-in. nominal thickness; (2) laminated planks of minimum 3-in. nominal thickness, set on edge close together and laid as required for floors; (3) approved 1⅛-in.-thick seven-ply tongue-and-groove interior grade plywood, bonded with adhesives identical to those used for exterior-type grades of plywood and meeting the requirements of U.S. Product Standard PSI-66, with all end joints staggered and butting on centers of beams spaced not over 4 ft apart or less; or (4) other deck material, if noncombustible.
Ordinary	No specific requirements
Wood frame	No specific requirements

NINE

Wind Uplift

Dynamic wind forces that produce uplift pressures are a major roof problem, especially in the tropical hurricane belt stretching from the Texas Gulf Coast northeastward through North Carolina.

Although this hurricane belt covers about 5 percent of continental United States, it accounts for more than one-third the total wind damage, proportionately about 10 times as much damage as occurs in other areas of the United States. *Tornadoes* occur over a much wider area, chiefly in the Southern, Southwestern, and Prairie states. Yet though they cause much greater damage, tornadoes account for less total wind damage than the more widely ranging hurricanes, for they strike much smaller areas.

Roofs can be practicably designed for anchorage to resist hurricane winds, which reach velocities around 150 mph. But for roofs on buildings caught in a tornado vortex, in which wind speeds may reach 500 mph, there is no practicable design solution. Buildings enveloped in a tornado vortex sometimes explode under the pressure differential created by the sudden atmospheric pressure drop.

Wind uplift stresses all the interfaces of the roof system—between framing members and deck, deck and vapor barrier, vapor barrier and insulation, and insulation and membrane.   To anchor these laminations the designer has these basic choices:

- Nailing, or other mechanical fastening
- Hot bituminous mopping
- Cold adhesives

See Table 9-1 for methods of anchoring roof system components.

The same organizations that set national fire-resistance standards, Underwriters' Laboratories, Inc. (UL) and Factory Mutual Engineering Corporation and Association (FM), have established tests and ratings for wind resistance.

MECHANICS OF WIND RESISTANCE

Against the windward wall that stops its natural, lateral movement, wind exerts a normal pressure dependent on its velocity (Fig. 9-1).

A building in a moving air mass behaves like an ungainly airfoil.   A level roof is analogous to the upper surface of an airplane wing, which through the negative pressure produced by the fast-flowing air, creates an uplift force that keeps the plane airborne.   Unlike an airplane wing, however, a sharply angled building is not designed for smooth, laminar flow.   At the windward roof edge, uplift forces developed across a roof trace a varying profile of -1.0 to -0.8 times the static wind pressure. Over the remaining roof area, uplift varies from -0.2 to -0.8 times the

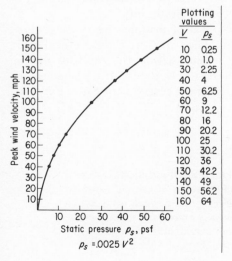

Plotting values	
V	p_s
10	0.25
20	1.0
30	2.25
40	4
50	6.25
60	9
70	12.2
80	16
90	20.2
100	25
110	30.2
120	36
130	42.2
140	49
150	56.2
160	64

$p_s = .0025 V^2$

FIG. 9-1   *Wind-velocity-to-pressure conversion chart.*

static pressure. At the windward corners, the uplift pressure may be three times the static pressure (see Fig. 9-2).

Additional uplift can result from air entering the building through broken or open windows or doors in the windward wall, thus increasing positive pressure inside.

The uplift forces depend basically on wind angle and roof slope. A wind blowing perpendicular to a plane wall produces maximum uplift. A level roof also produces maximum uplift. As roof slope increases, the suction on the windward roof plane decreases to zero at about 30 degrees. On steeper slopes, the wind exerts a positive pressure on windward roof planes and suction on leeward planes. Wind blowing perpendicular to a gable end wall produces suction over both sloping roof planes.

Less important factors affecting the wind forces on roofs are height, size, and shape. Because wind velocity increases with elevation, the roofs of tall buildings are subjected to greater uplift forces than low buildings. Since wind gusts may engulf a small building, while only partially engulfing a large building, wind-uplift forces are generally inversely proportional to building size.

Parapets introduce still another complicating factor. Tall parapets, 5 ft or so, can drastically reduce the maximum wind-uplift pressure. But a low parapet, of 18 in. or less, may increase the uplift pressure.

Uplift forces for testing wind resistance of roofing assemblies range up to 105-psf total uplift in the UL Bulletin of Research No. 52, "Development of Apparatus and Test Method for Determining Wind-Uplift Resistance of Roof Assemblies." A 105-psf uplift corresponds to a final design wind speed of nearly 180 mph. This is well beyond the highest wind velocity ever recorded in any United States city.

Hurricane wind-uplift forces, a reversal of loading two or even three times the normal gravity roof live load, may rip away the structural deck. Most wind-uplift failures, however, occur within the built-up roofing system, where the weakest plane is normally at the interface between the deck and the insulation or roofing membrane placed above it. For roofs

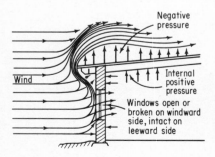

FIG. 9-2 *Wind uplift is especially severe at the roof perimeter, where it may equal, or at corners greatly exceed, the normal static pressure against the wall.* (*Factory Mutual System.*)

table 9-1   GENERAL ANCHORAGE METHODS

Key: F = Mechanical fasteners[a]
A = Adhesives[b]

Component to be applied	Wood (sawed plywood)	Pre-formed wood fiber	Gypsum (poured, precast)	Metal	Concrete (poured, precast)	Asbestos cement
Vapor barrier:						
Felt type..............	F	F	F	A	A	A
Plastic sheet...........	F-A	....	......	F[c]-A		
Metal foil.............	F-A					
Kraft paper...........	F-A	F-A	F-A	A	A	A
Insulation:						
Mineral-aggregate board	F-A	F-A	F-A[e]	F-A	A	A
Vegetable-fiber board...	F-A	F-A	F-A[e]	F-A	A	A
Glass-fiber board.......	A	A	F-A[e]	F-A	A	A
Glass foam board.......	A	A	F A[e]	F-A	A	A
Plastic foam...........	A	A	F-A[e]	F-A	A	A
Corkboard.............	A	A	F-A[e]	F-A	A	A
Lightweight concrete...	*Poured in place*					
Built-up membrane.......	F[d]	F[d]	F[d]	......	A	A

Deck substrate

Insulation:	Felt type	Plastic sheet	Metal foil	Kraft paper
Mineral-aggregate board	A	F-A	F-A	A
Vegetable-fiber board...	A	F-A	F-A	A
Glass-fiber board.......	A	F-A	F-A	A
Glass foam board.......	A	A	A	A
Plastic foam...........	A	A	A	A
Corkboard.............	A	A	A	A
Lightweight concrete...				

Vapor-barrier substrate

	Mineral-aggregate board	Vegetable-fiber board	Glass-fiber board	Glass foam board	Plastic foam	Corkboard	Lightweight concrete
Base sheet..............	A	A	A	A	A	A	F-A

Insulation Substrate

[a] Mechanical fasteners are nails or special fasteners.
[b] Adhesives are hot bitumen or fire-rated, cold-applied mixtures.
[c] Fastened through insulation.
[d] Base ply (or plies) only, unless slope requires back nailing.
[e] Apply coated sheet on gypsum substrate before installing insulation board.

with prefabricated insulation, the insulation-membrane interface is sometimes the weakest plane; the wind may peel the membrane off the insulation, which remains in place on top of the deck. More often, bond failure occurs between deck and insulation. Sometimes both interfaces fail.

CAUSES OF ROOF BLOWOFFS

Most roof blowoffs start with failure at the edge flashing or gravel stop. According to 90 percent of FM inspectors' reports on wind-uplift failures, an initial lifting of the fascia piece exposes the roof to a progressive bond failure after wind penetrates the edge, deforms the edge flashing, and starts peeling the roof covering.

Some of these edge-flashing failures stem from poor design and detailing. Heavy sheet metal (at least 18 gage) should be specified to resist wind distortion. Metal fascia strips should be nailed to blocking at maximum 3-in. spacing. Edge flashing should be detailed to prevent wind entry into steel deck flutes at the upper deck surface directly under the insulation, where it adds positive pressure to the negative pressure at the surface. Faulty installation—failure to nail at a specified spacing, distortion of flashing metal, and other field errors or abuse—all contribute to uplift failures attributable to faulty edge flashing. Omission of an edge nailer resulted in the wind failure shown in Fig. 9-3.

The common practice of leaving roof coverings incomplete for several feet back from the roof edge, with the edge flashing and closing membrane installation performed later, can cause a wind-uplift blowoff during construction. This practice allows the wind to get under the membrane edge and peel back the covering. When it is unavoidable, the incomplete

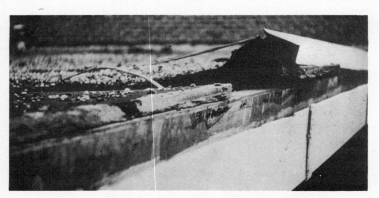

FIG. 9-3 *Wind forces ripped this metal gravel stop from its insecure connection to the insulation—a costly lesson on the need for anchoring to wood nailers around the perimeter of a nonnailable deck. The nailer also serves as an insulation stop. (GAF Corpration.)*

FIG. 9-4   *Random pattern of cold adhesive on this wind-peeled steel deck contains an obvious clue to the failure.   The adhesive ribbons should run straight and* parallel *to the flutes at maximum 6-in. spacing, not in the widely spaced, random pattern running* across *the flutes.   (Factory Mutual System.)*

membrane edge should be temporarily lined with weights until it is spliced.

Other factors in wind-uplift failure are:

- Faulty practice in installing cold-applied adhesives
- Excessive longitudinal or transverse deflection (dishing) of lightgage steel decks with cold-applied adhesives
- Cold-weather application of hot-mopped bituminous adhesive
- Faulty application, or spacing, of mechanical fasteners

The wind-uplift failure shown in Fig. 9-4 shows a grossly faulty adhesive pattern, across instead of parallel to the deck flutes.   The adhesive was applied in a random, overspaced pattern instead of the specified 6-in. spacing.

Because they are so flexible, the thinner metal roof-deck sections—24 or 26 gage—can cause serious trouble for cold-applied adhesives. Transverse deflection (dishing) of light, springy decks can break contact between the steel surface and the rigid insulation above, thus precluding good bond.   Some roofers, confronted with a specification calling for cold adhesive, have recommended and obtained authorization to substitute mechanical fasteners.   To reduce the risk of excessive steel deck deflection, FM Engineering Associates recommend the use of a minimum 22-gage steel deck.

Cold-weather application of hot-mopped bitumen increases the hazard of roof blowoffs.   Cooled bitumen will not tightly bond the insulation to the deck, or the membrane to the insulation.   Factory Mutual is investigating roof performance in wind in relation to application temperature.

COMPUTING WIND UPLIFT

To compute wind uplift on roofs of buildings 50 ft or less in height, proceed as follows:

1. Obtain the fastest-mile speed recorded by the local U.S. Weather Bureau Station or use Fig. 9-5 and Table 9-2.   (The fastest-mile wind speed in the highest recorded speed attained by wind traveling 1 mile.)

2. Multiply the fastest-mile wind speed by a gust factor.   For buildings less than 130 ft in longest dimension, the gust factor is 1.3; for buildings whose *shortest* horizontal dimension exceeds 130 ft, the gust factor is 1.1 (for intermediate buildings, use a gust factor of 1.2).

3. Get the static pressure from Fig. 9-1 or compute from the following formula:

$$p_s = 0.0025V^2$$

where p_s = static pressure (psf) and V = wind speed (mph).

4. Multiply the static pressure by appropriate shape factor.   For low industrial buildings with roof slopes less than 20 degrees, use a shape factor of 1.5.   (Buildings of greater roof slope or unusual shape require more elaborate investigation.)

EXAMPLE: Determine the wind uplift for a low, level-roofed building, 200 × 400 ft in plan, in Des Moines, Iowa.

1. Design wind speed = 80.5 (fastest-mile wind speed)
2. Peak wind speed (corrected for gust factor) 1.1 × 80.5 = 88 mph
3. Static pressure (from graph) = 20 psf
4. Static pressure corrected for shape factor = 1.5 × 20 = 30 psf

A 50 percent uplift reduction is allowable for windowless buildings with few exterior doors normally kept closed.   For such a building, the design wind-uplift pressure in the above example would be 15 psf.

For the central area of large roofs, design uplift pressure can be reduced by 20 percent.   The reduced-pressure area must be bounded by lines at least four times the building height in from the building edge.   If the Des Moines building is 20 ft high, this reduced-pressure area, designed for 24 psf uplift, could consist of a strip 40 ft wide and 240 ft long.

A basic design uplift pressure of 30 psf [see (4) in above example] is the minimum recommended by FM engineers, even if the maximum fastest-mile wind speed recorded by the local U.S. Weather Bureau indicates a lower acceptable value.

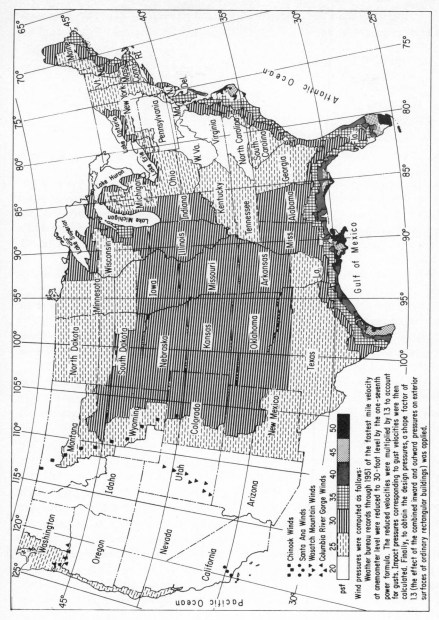

Wind pressures were computed as follows:
Weather bureau records through 1951 of the fastest mile velocity
at anemometer level were reduced to 30-foot level by the one-seventh
power formula. The reduced velocities were multiplied by 1.3 to account
for gusts. Impact pressures corresponding to gust velocities were then
calculated. Finally, to obtain the design pressures, a shape factor of
1.3 (the effect of the combined inward and outward pressures on exterior
surfaces of ordinary rectangular buildings) was applied.

■ ■ Chinook Winds
▾ ▾ Santa Ana Winds
▾ ▾ Wasatch Mountain Winds
▲ ▲ Columbia River Gorge Winds

psf | 20 | 25 | 30 | 35 | 40 | 45 | 50 |

FIG. 9-5   *Minimum allowable resultant wind pressures.*   (*American National Standards Institute.*)

For buildings over 50 ft high, design uplift pressures should be increased by the following factors:

Height, ft
50–99	1.25
100–999	1.50
500–1199	1.75
1,200 plus	2.0

ANCHORING TECHNIQUES

The Livonia fire brought drastic changes in the anchoring of combustible insulation to steel decks; it stimulated research into new cold-applied adhesives and mechanical fasteners to replace the hazardous, fire-feeding bitumens. In addition to fire requirements, a steel-deck roof assembly must withstand a 60-psf suction for 1 min to qualify for the FM Class 1 rating.

Wind tests for other deck materials are largely random; a great many roof assemblies remain untested. Underwriters' Laboratories publishes a list of roof-deck assemblies rated class 30, 60, or 90, depending on their successful resistance to 45-, 75-, or 105-psf total negative pressure in the UL uplift test. (See UL "Building Materials List," Roof Deck Constructions. FM also rates deck materials other than steel decks. These are published in the FM "List of Approved Materials.")

Approved anchoring techniques satisfying both the fire-spread and wind-uplift requirements for Class I insulated steel deck assemblies are listed in FM Engineering Associates' "Loss Prevention Data Sheet 1-28S."

Factory Mutual has approved a number of proprietary cold-applied adhesives for anchoring insulation or vapor barriers to Class I steel decks. Cold-applied adhesives include clay-filled bituminous emulsions that will neither burn nor drip, chlorinated, rubber-based solvents, and other liquid compounds with combustible vapors that dissipate as the adhesive dries. These adhesives are generally applied in parallel ribbons about $\frac{1}{2}$ in. wide, with 6-in. spacing, in quantities listed in the manufacturers catalogues and in FM tables (see Fig. 9-6). When the insulation or vapor-barrier sheet is pressed into place, the flattened ribbons spread to $1\frac{1}{2}$ or 2 in. wide. On steel decks with flutes more than 1 in. in width, these ribbons must run parallel, never perpendicular, to the ribs (see Fig. 9-6).

For noncombustible insulation, e.g., foamed glass, glass fiber, preformed perlite board, Class I steel decks may have hot-mopped bituminous adhesive, which provides excellent resistance to wind uplift.

FIG. 9-6 *Cold-applied adhesive, spread in narrow ribbons, qualifies steel-deck assembly with board insulation for Class I rating (left). Locking mechanical fastener, driven with rubber-headed mallet through plastic vapor-barrier sheet and insulation board (right), provides alternative anchorage method for Class I fire-rated steel deck. (Insulation can be bonded to vapor barrier with ribbons of cold-applied adhesive instead of mechanical fasteners.)*

There are only two FM-approved mechanical fasteners for Class I steel-deck assemblies. One has a locking tongue that springs open after the fastener point penetrates the steel deck; the other has a serrated shank that grips the penetrated steel (see Fig. 5-4). Both these fasteners may be used with vapor barriers sandwiched between deck and insulation.

Other mechanical fasteners supplement the stock of traditional large-headed roofing nails still used on wood plank decks. For low-density materials (e.g., insulating concrete, preformed wood fiber, or gypsum), one manufacturer's fastener has a collapsible head driven against a cap located farther down the shank. As the head collapses, a locking foot rotates out from the shank to anchor the fastener against uplift. Another nail, designed especially for insulating concrete, has a tapered tubular shank that expands on driving to develop frictional resistance to uplift (see Fig. 5-4).

For nails and other mechanical fasteners, the Asphalt Roofing Manufacturers' Association Committee on Built-up Roofing recommends a minimum 40-lb pullout strength. One nailed plywood roof deck assembly has qualified for UL Wind Uplift Class 60. UL Construction NM501A requires 2-in.-wide masking tape over all plywood joints, to prevent air leakage from the underside. Nails are ring shank type, with

1-in.-square washers.   Maximum spacing is 12 in. on centers, in any direction.

The goal of better wind-uplift resistance is another reason for completing insulation and membrane installation in one day's operation, rather than delaying application of the whole membrane or its top plies. The uplift resistance of a nailed deck is greatly increased after mopping of the top plies.   The mopped plies stiffen the whole membrane and thereby equalize stresses instead of permitting stress concentrations to start progressive failure through large areas of the roof.

The determination of nail pullout strength is fairly straightforward for wood and preformed wood fiber (see Fig. 9-7).   But for poured-in-place materials like gypsum and insulating concrete, it is more complex.

The pullout strength of nails anchored in insulating concrete (perlite, vermiculite) depends chiefly on the mix ratio and ultimate 28-day strength and the deck's age when the nail is pulled.   A proprietary nail, with a tubular shank tapering upward to the cap, has been tested at values consistently exceeding 100 lb for both vermiculite and perlite concrete of 1:4 mix cured for 7 days or more.

BUILDING CODE REQUIREMENTS

The widely used American National Standards Institute (ANSI) "Building Code Requirements for Minimum Design Loads in Buildings and Other Structures" (ANSI A58.1-1955) discriminates among different sections of continental United States.   Under ANSI requirements, design

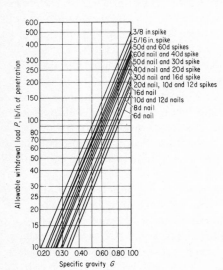

FIG. 9-7   *Allowable withdrawal loads for nails and spikes in pounds per inch of penetration, for one nail or spike installed in side grain under normal duration of loading.   (American Institute of Timber Construction.)*

uplift pressures go up to 125 psf for a hypothetical super skyscraper (over 1,200 ft high) in an extreme hurricane zone (e.g., Key West, Florida).

To find the correct column for determining wind-uplift pressure under the ANSI code, the designer first consults the wind-pressure map, Fig. 9-5, for the basic wind pressure, in psf, at 30 to 49 ft above grade. He then enters the proper column in Table 9-2 and the proper roof-height row. (Roof height is mean height above mean grade level around the building.) Thus for Miami, Florida, which lies in the 45-psf area, the roof of a 400-ft-high office tower would be designed for 88-psf uplift.

table 9-2   WIND-UPLIFT PRESSURES FOR VARIOUS
ZONES ABOVE GROUND

(Adapted from ANSI "Building Code
Requirements for Minimum Design Loads
in Buildings and Other Structures,"
ANSI A58.1-1955, Section 5.3.1)

Height zone, ft	Wind-pressure–map area psf*						
	20	25	30	35	40	45	50
Less than 30	19	25	31	31	38	44	50
30–49	25	31	38	44	50	56	62
50–99	31	38	50	56	62	69	75
100–499	38	50	56	69	75	88	94
500–1,199	44	56	69	75	88	100	112
1,200 and higher	50	62	75	88	100	112	125

* For wind-pressure–map area see Fig. 9-5, "Minimum Allowable Resultant Wind Pressures."

The ANSI requirements form the basis for three of the four national model building code wind-load provisions. The "National Building Code," the "Uniform Building Code," and the "Southern Standard Building Code" contain the ANSI wind-uplift loadings with only slight, individual modifications. The BOCA "Basic Building Code" has similar provisions, differing mainly in wind-height zones and uplift factor.

WIND ALERTS

General

1. Before specifying a roof assembly, consult with the local building official and insurance agent or rating bureau to make sure that your proposed design satisfies local wind requirements.

2. Before specifying an anchoring technique:
    a. Request data on wind-uplift resistance from manufacturer.
    b. Check UL or FM listings for wind-uplift approval of proprietary products.

Design

1. Beware of substituting even a single component within a roof assembly rated for wind uplift. Such changes require restudy of the resulting new roof assembly. Wind-uplift ratings apply to an entire roof assembly, not to its individual components.

2. Regardless of more lenient code requirements, design roofs to resist minimum 30-lb uplift, except for (a) roofs of windowless buildings; (b) interior areas of large roofs.

3. In hurricane zones, double the number of nails or other mechanical fasteners required per square throughout an area extending 10 ft in from the roof edge.

4. Specify minimum 22-gage thickness for steel decks.

5. Specify minimum 18-gage for metal gravel stops or fascia strips. Require maximum 3- or 4-in. nail spacing for all such edge flashing.

6. Specify installation of cold-applied adhesives in ribbons 6 in. on centers *parallel* to the ribs of metal decks with flutes more than 1 in. wide.

Field

1. Require weighting of all membrane edges left incomplete before splicing with other sections of membrane.

2. Before application of cold adhesive on metal deck, inspect for permanent deflections that could prevent or break bonding contact with insulation. Reject a deck that deviates from a plane transverse surface more than $\frac{1}{16}$ in. between adjacent ribs.

TEN

Specifications and Performance Criteria

SPECIFICATIONS AND DRAWINGS

The specifications for the roof system should define and establish the materials to be used, the installation techniques, the limits on moisture, temperature, and other environmental conditions permitted during application, and the apportionment of responsibilities. Because the written word is more easily understood by lawyers and other laymen, specifications usually carry greater legal weight than drawings.

Ideally, the specifications should amplify, but not repeat, information on the drawings. The architect's drawings graphically convey information on design details, location, and dimensions. The specifications establish standards for material quality and workmanship.

Drawing Requirements

The drawings (including schedules) should cover the following information:

1. Scope—location and types of roofing, change of roof levels on plans, elevations and details
2. Slopes and drains
3. Walkways
4. Insulation location and thickness, lateral ventilation of insulation, stack vents
5. Details of cants, openings, crickets, curbs, expansion joints, eaves, and stack and edge vents
6. Ventilators, skylights, scuttles, and equipment
7. Wood nailers to receive flashing (both composition and metal)

Specification Requirements

The specifications should provide the following information:
1. Scope of the work should combine roofing membrane, roofing sheet metal, accessories, insulation, and vapor barrier.
2. Selection of a roofing system compatible with a reputable manufacturer's published specification.
3. Coordination of roofing specification with other relevant specifications.
4. Responsibilities of the different members of the construction team—owner, architect, manufacturer, general contractor, roofing and sheet-metal contractor, mechanical contractor, plumber.
5. Description of vapor barrier, insulation, felt, bitumen, and flashing materials, with tests and standards for acceptance.
6. Maximum and minimum roof slopes (coordinated with deck specifications and structural drawings).
7. Methods of applying all components, with limiting conditions, e.g., heating of bitumen.
8. Provisions for inspection, requirements for bitumen, anchorage requirements for vapor barrier, insulation, and membrane.
9. References to (but not repetition of) mandatory provisions already stated in the general conditions.

WRITING STYLE

Roofing specifications must be written in clear, concise prose.   Here are several rules:
1. Don't parrot legal jargon ("said" and "same" as identifying pronouns) or long-winded, all-purpose safety clauses copied from old specifications.   Cluttering a specification with needless words obscures the essential information and reduces the chances that the specifications will be read.
2. Avoid indefinite expressions, e.g., "reasonable," "best quality," "or equal."

3. Don't use synonyms; repeat the same word. (Choose, for example, "built-up roofing," "membrane," or "built-up roofing membrane" and retain it throughout.)

4. Give directions, not suggestions.

DIVISION OF RESPONSIBILITY

In assigning responsibility, the specification writer should follow these general principles:

1. The *architect* bears ultimate responsibility for the design of all components in the built-up roof system and for its final inspection and acceptance. He is ultimately responsible for use of a specified material and its compatibility with adjoining materials, regardless of information received from manufacturers, suppliers, or applicators.

2. The *manufacturer* (or manufacturers) of the various roofing components is responsible for furnishing materials that conform with the specifications and for furnishing accurate technical data on the physical properties, chemical composition, and other information pertaining to his material and its compatibility with adjoining materials.

3. The *general contractor* is responsible for:

   a. Coordinating the work of the roofing contractor with that of other subcontractors penetrating, adjoining, or working on the roof
   b. Insuring that the roofing subcontractor follows the specifications
   c. Insuring the protection of stored roofing materials and components from moisture and other hazards before and after installation
   d. Providing walkways, where required, to protect the roof from traffic damage
   e. Installing a satisfactory deck
   f. Insuring that all roof-penetrating elements and perimeter walls are in place, and that perimeter nailing strips, cants, curbs, and similar accessories are installed *before roofing work starts*

Particularly in the use of new roofing system products, the architect should follow the principles of the AIA "Policy Statement on Building Product Development and Uses," which is reprinted below in full.

POLICY STATEMENT ON BUILDING PRODUCT DEVELOPMENT AND USES[1]

PRINCIPLES

The development and introduction of new products is essential to the growth of the building industry. Equally important is a proper understanding of

[1] Reprints are available from The American Institute of Architects, 1735 New York Avenue, N.W., Washington, D.C., 20006.

the characteristics, properties, and appropriate uses of both new and previously existing products. Assuring the successful use and application of products as well as their related assemblies greatly minimizes the possibility, during construction and after completion, of failures which have frequently caused costly delays and grievous loss to all connected with the project.

The purpose of this policy statement is to encourage continued development of new building products and better uses for existing ones, to add to the existing technical knowledge concerning these products, and to foster better understanding between the parties involved in the building process, all being conducive to improved building design and accelerated technological progress in the entire building industry.

DEFINITION OF TERMS

The following explanatory material is for clarification of this policy statement and is not intended to be all-inclusive.

Manufacturer refers to the producer of the product, system, or assembly, and includes, in addition to the prime supplier, his sales and distribution outlets and their respective representatives. The term does not include contractors independent of the organization of the original producer.

Contractor refers to independent prime contractors, subcontractors, and their employees installing specific building products.

Architect refers to the individual or organization as defined in the AIA documents, engaged in the design of construction and buildings, and includes engineers and others in the architect's employ or under his control.

OBLIGATIONS OF THE PARTIES

Manufacturer: The Manufacturer should supply the Architect with all essential data concerning his product, including pertinent information which would involve its installation, use, and maintenance. In addition to the physical properties, this should take the form of chemical descriptions, laboratory and field test results, standards and ratings.

Particularly important is information on the product's compatibility and interfitting with interrelated products, as well as precautions and specific warnings on where the product should not be used based on conditions of known or anticipated failures.

The Manufacturer is expected to supply pertinent data concerning the compatibility, physical relationship, and maintenance of his product when combined with other products. Whenever the Manufacturer has specific knowledge of an improper use of his product, he should furnish such information in writing to the Architect. It is recommended that this type of information and communication emanate from the manufacturer's technical staffs, such as application engineering and researching divisions. It is further recommended that the services of public relations and advertising counsel be reserved primarily for improvement in presentation and format of product literature.

When an inquiry is made by an Architect of the Manufacturer concerning an intended use and installation of his product, the Manufacturer is expected to respond with a technically qualified reply.

The Manufacturer is expected to recognize that he is responsible for the failure of his Product to perform in accordance with written data supplied by him or his authorized representatives, as well as misrepresentations of such data.

When a Product has been installed in accordance with the Manufacturer's written instructions and written recommendations, and such Product fails, then the Manufacturer has the responsibility therefor. Such responsibility extends to related products affected by the failure, where the Manufacturer has notice of the proposed use of such related products. In case of failure, manufacturers of the other products involved should make available the technical knowledge to the Architect in the correction of the failure.

The Manufacturer is expected to investigate the relation of his product to other components likely or logically expected to be used in association with his. Such information should be available to the Architect.

Architect: The Architect is responsible for proper design. He is expected to inform himself with respect to the properties of the products he specifies, through he is entitled to rely on Manufacturers' written representations. He is advised to seek the technical opinion of the Research or Application Engineering Departments of the Manufacturer when his intended use is not clearly included in the printed data of the Manufacturer. He is further responsible for uses contrary to supplementary written information on proper use and installation procedures of the Manufacturer.

The Architect's use of a Product and its installation should extend to its compatibility with and relationship to adjacent materials and assemblies, notwithstanding the Manufacturer's similar obligations.

Contractor: It is the responsibility of the Contractor to inform himself concerning the application of the Products he uses and to follow the directions of the Architect and Manufacturer.

In the event of disagreement between the Contract documents and the Manufacturer's directions, the Contractor is expected to seek written instructions from the Architect before proceeding with the installation.

If the Contractor has knowledge of or reason to believe the likelihood of failure, he is expected to transmit such knowledge to the Architect and ask for written instructions before proceeding with the work.

Owner: It is assumed that the Owner or other person responsible for operation and maintenance of the project will properly maintain the material and equipment in accordance with Manufacturer's recommendations.

SPECIFICATION ALERTS

General

1. Specify roofing materials and components preferably to conform with ASTM standards or federal specifications as a minimum, with any additional requirements appended to the selected standards. Use of

standards from a single source simplifies the task of checking specifications and promotes consistency.   Make sure that the specifications are understood and are available for reference.

2. Avoid duplication of material properties covered by specifications, e.g., "aggregate shall be ¼ to ⅝ in. in size, clean, and free from dust and foreign matter."   Instead, require conformance with ASTM Specification D1863-64.

3. Check other specification sections for inclusion of information necessary for other roof system components.   Example:

Section 0610—Rough Carpentry:

- Include wood nailers to secure metal flanges and devices requiring connection with bituminous flashing.
- Include openings in wood members at edges of insulation to allow venting, if not provided by the design of the insulation.

4. Require manufacturer of flashing materials to approve compatibility with roofing materials in coefficient of thermal expansion, adhesion, elasticity, resistance to sunlight, and other relevant properties.

General Field Alerts

1. Require a joint job inspection and conference by the roofing contractor, general contractor, architect's representative, manufacturer's representative, and building inspector before application starts.

2. Require the general contractor to have all drains, curbs, cants, perimeter walls, and roof-penetrating building components in place *before roofing work starts*.   (Installation of these elements after roofing work starts will require needless patching and repairs that will multiply the chances for leaks and other modes of premature failure.)

3. Require the roofing contractor to approve the substrate surface as satisfactory *before* he starts work.

4. Require the general contractor to give the architect ample notice of intent to start roofing operations (usually two days or more).

5. Require the general contractor to coordinate the work of the roofing subcontractor, mechanical subcontractor, or other subcontractors, with each assuming full responsibility for any damage he may inflict on another's work.

6. Require subcontractors installing rooftop equipment to notify the architect and general contractor of any damage they cause to the mem-

brane. Require offending subcontractors to finance the necessary repairs made by the roofing subcontractor.

Technical Alerts

For technical recommendations for the various roofing components, consult the Alerts in the following chapters: Chap. 3, "Structural Deck," Chap. 4, "Vapor Control," Chap. 5, "Thermal Insulation," Chap. 6, "Element of the Built-up Roofing Membrane," Chap. 7, "Flashings and Accessories," Chap. 8, "Fire Resistance," and Chap. 9, "Wind Uplift." As a guide to writing the specification, use the specification work sheet that follows.

SPECIFICATION WORK SHEET FOR BUILT-UP ROOFING SYSTEMS

The following specification work sheet is neither a "master" specification that enumerates all possible options nor a "canned" specification that can, in some instances, be reprinted. It is intended only as a time-saving specification checklist, to which unusual requirements can be added. Information from this manual will be incorporated into pertinent sections of a master specification by Production Systems for Architects and Engineers, Inc. (PSAE), 343 South Dearborn Street, Chicago, Illinois 60604. PSAE was sponsored by The American Institute of Architects to provide a national computerized master specification system available for the design professions.

In using the work sheet, the designer chooses from words, phrases, numerals, or blanks, which are separated by dashes or brackets, as follows:

▪ *Dashes* (—,). Dashes separating a series of two or more items indicate that one or more of the included choices may be selected or marked out, and that blanks may be filled as required to conform with the specific project.

▪ *Brackets* [ ]. Brackets have several uses:

1. To set apart choices in statements within which internal choices must also be made

2. To set apart optional statements covering unusual conditions or more detailed descriptions that may be needed for some projects

3. To set apart optional statements adding specific or complementary information to previously covered material

BUILT-UP BITUMINOUS ROOFING
INCLUDING
PREPARED ROLL ROOFING

Requirements of the Conditions of the Contract, and of Division 1 of these Specifications apply to Work under this Section.   *Note 1*

01 SCOPE OF THIS SECTION:

A. FURNISH AND INSTALL: *Note 2*
  1. ....   *Note 3*
  2. ....

B. FURNISH UNDER THIS SECTION FOR INSTALLATION—UNDER   *Note 4*
   OTHER SECTIONS OF THESE SPECIFICATIONS—BY OTHERS
   NOT PARTY TO THIS CONTRACT—:
  1. ....
  2. ....

C. FURNISHED—UNDER OTHER SECTIONS OF THESE SPECI-   *Note 5*
   FICATIONS—BY OTHERS NOT PARTY TO THIS CONTRACT—
   FOR INSTALLATION UNDER THIS SECTION:
  1. ....
  2. ....

D. NOT INCLUDED: *Note 6*
  1. ....
  2. ....

02 GENERAL PROVISIONS:

A. NOTES: *Note 7*
  1. *Delivery And Storage:*
    a. Materials:
      ▪ Deliver materials with manufacturers' labels intact,   *Note 8*
        and legible.
      ▪ Store—roofing insulation—roofing felts—roofing
        fabrics—..................—indoors—on raised platforms
        and covered with—suitable weatherproof protective
        coverings—such as—tarpaulins—plastic sheeting—
        ....—. Provide continuous protection
        against wetting of materials. Remove wet mate-
        rials from Project site.

      ▪ ....   *Note 9*
    b. ....
  2. ....

B. PRIOR APPROVALS: Before fabrication, delivery, or installa-   *Note 9*
   tion of the Work, submit and obtain approval of the following in accordance with requirements of—Division 1—............
   ....—except as modified herein:
  1. *Applicator* (Roofing Subcontractor):
    a. Letter from roofing materials' manufacturer approving   *Note 10*
       roofing subcontractor.
    b. ....

2. *Manufacturer:*
   a. Letter from manufacturer stating that he will bond   *Note 10*
      roof as detailed and specified.
   b. Letter from manufacturer stating specification pro-   *Note 10*
      posed for use in providing for requirements of Contract
      Documents. Include a copy of latest edition of
      manufacturer's proposed specification.
   c. ....

3. *Materials Proposed:* with Trade Names and   *Note 11*
   Manufacturer.
4. *Samples:*—in triplicate—.....................—as requested.
5. *Regulating Agency Approvals:*
   a. Submit—..................... copies of test report of fire test
      from—UL—FM—, ... on roofing system—
      ....—.
   b. Submit—..................... copies of evidence or—UL or FM
      Listing for classification required for specific locations
      in Project—...
   c. Submit—................ copies of evidence of Fire Hazard
      Classification as listed by UL or FM for specific
      products required for Project.
   d. ....

6. ....

C. QUALITY CONTROL:
   1. *Certificates:*   *Note 12*
      a. Submit in—duplicate—.....................—a certificate
         from the manufacturer, properly attested, stating that
         —bulk material—non-labeled material—.....................
         ....— complies with the requirements of the
         Contract Documents. Furnish certificate prior to—
         installation—...—.
      b. ....
   2. *Tests:* Examine test-cut specimens in accordance with
      recommended procedure of the National Roofing Con-
      tractors Association.
      a. Before installing—flood coat and surfacing aggregate
         —...—,
         cut—4—.....................—samples of built-up roofing—as
         required by Architect—...—.
         [Samples shall be approximately—4 in. by 3 ft 4 in.—
         ....—, —taken perpendicular to
         long dimension of felts—..
         ....—.]

   3. ....

D. INDUSTRY STANDARDS:
   1. Work specified by reference to the published standards or
      specifications of a manufacturer or organization shall
      comply with the requirements of—the current standard

or specification listed—the standard or specification
listed (edition with date listed)—.

  a. In case of conflict between the referenced specification
and the project specifications, the project specifications shall govern.

  b. In case of conflicts between the referenced specifications or standards, the one having the more stringent
requirements shall govern.

  c. ....

2. Specifications or standards of the following are referred to
in this Section by the accompanying abbreviations:

    American Society for Testing and Materials (ASTM)
    Underwriters' Laboratories, Inc. (UL)
    American National Standards Institute (ANSI)
    Factory Mutual Engineering Corporation and
    Association (FM)

3. ....

E. ....

03 MATERIALS:

A. ACCEPTABLE MANUFACTURERS: *Note 14*

1. Except as provided for herein,—materials—...................—
shall be the products of one of the following
manufacturers:

  a. ....
  b. ....

2. ....

B. SUBSTITUTIONS: *Note 15*

1. In addition to requirements of—Conditions of the
Contract—Division 1—.............................—for submittals for use by Architect in considering requested
substitutions, submit the following:

  a. ....
  b. ....

[2. Submit requests for, and obtain approval of substitutions
prior to purchase and delivery.]

3. Include requested substitutions (after approval) in lists
submitted for approval under "Prior Approval."

4. ....

C. MATERIALS LIST: *Note 14*

1. *Built-up Roofing:*   *Note 16*

[a. Roofing system shall be selected from standard
specifications of—one of the manufacturers listed
hereinbefore—an acceptable roofing manufacturer—
............................—and shall be designed to meet
the following conditions: *Note 17*

- Deck: _____
- Vapor Barrier: _____   *Note 30*
- Deck Slope: _____
- Insulation: _____   *Note 32*
- Base Sheet: _____
- Roofing Bitumen:_____   *Note 34*
- Roofing Felts:_____
- Roofing Fabrics: _____
- Roofing Surfacing: _____   *Note 35*
- Roofing—Design Life—Bond Period—_____   *Note 36*
  _____—:_____
- _____]

[b. Roofing system shall be a _____ year bonded type,—   *Note 23*
with _____ bitumen and _____
surfacing. The system shall be—selected from one of
the acceptable manufacturers'—one of the following
manufacturers'—specifications as listed: *Note 16*
- _____, Spec. No. _____
- _____, Spec. No. _____   *Note 28*
- _____, Spec. No. _____]

   c. _____

2. *Bitumens:*   *Note 18*
   a. *Primer:*
- Asphalt:—ASTM D 41—_____—_____—.   *Note 34*
- _____

   b. *Cementing Coats:* (Mopping Coat)
- Asphalt:—[ASTM D312-_____,—Type I—Type   *Note 34*
  _____—]—

- Coal-Tar Pitch:—ASTM D450-_____, Type A—   *Note 34*
  _____—

- Cold-Applied Roofing Asphalt:—Cut-back—   *Note 19*

- _____
  _____

   c. _____   *Note 35*

3. *Felts:*
   a. Asphalt-Saturated Roofing Felt:—ASTM D226-   *Note 18*
  _____—_____—.
   b. Asphalt-Saturated Asbestos Felt:—ASTM D250-   *Note 34*
  _____—_____—.
   c. Coal-Tar-Saturated Roofing Felt:—ASTM D227-
  _____—_____—.
   d. Asphalt-Impregnated Glass Fiber Mat (Felt)—ASTM
  D2178—
  _____—_____—.
   e. Base Sheet:—ASTM D2626—   *Note 20*
  _____—_____—.

      f. Cap Sheet: ....

      g. Prepared Roll Roofing:—ASTM D249—, —with   *Note* 21
         D371 19-in. selvage—..

      h. ....

4. *Fabric:*   *Note* 18

      a. Bituminous-Saturated Woven Cotton ASTM D173

      b. Bituminous-Saturated Woven Glass ASTM D1668

5. *Building Paper:*   *Note* 22

      a. Rosin Paper:—Rosin-sized—sheathing—.................—
         paper, weighing approximately—5 lb per 100 sq ft—

         ....

      b. ....

6. *Insulation:*

      a. Fiberboard:—.............in. thick—of thickness shown   *Note* 24
         on Drawings—, in sheet sizes of—2 ft by 4 ft.................
         ....—.
         [Insulation shall be treated to resist fungi;—shall be
         vapor-proofed on both faces and all edges—shall be
         asphalt-impregnated—..—]   *Note* 24

      b. Fibrous Glass:—.........in. thick—of thickness shown
         on Drawings—, in sheet sizes of—2 ft by 4 ft—.............
         ft by ....ft—.   *Note* 24
         [Insulation shall be "..."
         manufactured by..]
         [Insulation shall have a density of—12 lb per cu ft—
         ....lb per cu ft—and shall be covered—on one side
         with asphalt at the rate of 3 oz per sq ft; and shall be
         covered on the nonasphalted side, and around all
         edges with a heavy kraft paper—...............................]

      c. Cellular Glass:—............................as mfd. by...........
         ....—.......................
         ....—

      d. Volcanic-Glass:—..........................as mfd. by........
         ....—composed of.................
         ....—with compressive strength of
         ....psi at .... % consolidation—with Conductance
         Value not exceeding.........for nominal—1 in.—.............
         —thickness...—

      e. ....

7. *Cants—and Eaves—:*

      a. Preformed fiber—............................, —.................
         ....as mfd. by..—

      b. Eave strip shall be tapered from—1⅝ in.—.............
         in.—to.........in. in—1 ft 3 in.—....................—
         width.

      c. ....

8. *Plastic Cement:*
   a. ....—fibrated asphalt—as
      manufactured by...
      ....
   b. ....
9. *Mineral Aggregate Topping:*
   a. Conforming to—ASTM D1863-.........—.................—
   b. Crushed—stone—gravel—slag—...................................—
   c. ....
10. *Fasteners:*   *Note 25*
    a. Nails:
       ▪ Nails shall be of sufficient length to pass through   *Note 38*
         insulation, and shall make a maximum penetration
         into the deck without protruding—(except for
         metal decks)—.]
       ▪ Nails shall be—large headed—...................................—,
         —galvanized roofing nails—...................................—,
         —1 in.—.................—long,—No. 12 stub gauge—
         ....—]
       ▪ Nails shall be self-clinching, —...................................—
         " ...." as manufactured by...............
         ....—, ...long—.]
       ▪ ....
    b. Mechanical Fasteners:
       ▪ ....
    c. Fasteners shall be type, length, and finish as re-
       quired by the manufacturer of—insulation—roofing
       system—.
11. *Walkway & Protection Course:*
    a. ....as mfd. by.....................   *Note 37*
    b. Homogeneous core of asphalt, plasticizers, and inert
       fillers. Core bonded by heat and pressure between
       two saturated and coated sheets of organic felt.—
       [One surface coated with.........ceramic granules in
       ....color—...................................—]
       —...—.
       Size:—3 ft 0 in.—.........—by—6 ft 0 in.—.............—by—
       ½ in.—.................—thick.
    c. ....
12. ....

04 PERFORMANCE:

A. LOCATION: *Note 26*
   1. Dead-Level Asphalt: Over—Space(s)..................—.............
   2. Coal-Tar Pitch: Over—Space(s)....................—...............
   3. Steep Asphalt: Over—Space(s)....................—...............

4. Prepared Roll Roofing—Type............\\.........—: Over— Space(s)...........................—...
5. ....

B. PREPARATION: *Note 27*
   1. Surfaces to which roofing is to be applied shall be even, smooth, sound, thoroughly clean and dry, —..................... ....—, and free from defects that might affect the soundness of the installation.
   2. ....

C. TEMPORARY ROOFING: *Note 28*
   1. Provide temporary roofing—over.............................—.
   [2. Temporary roofing shall consist of.................................... ....]
   [3. Vapor barrier shall not be used as a temporary roofing.]
   4. Install temporary roofing in accordance with requirements of manufacturer of roofing system.
   5. ....

D. SHEATHING PAPER: *Note 22*
   1. Prior to application of—insulation—roofing felts—............ ....—on...deck, install— rosin-sized—........................—sheathing paper in accordance with requirements of roofing manufacturer's specification.
   2. ....

E. VAPOR BARRIER:
   1. Install vapor barrier on deck prior to installing insulation.   *Note 29*
   [2. Install one ply of coated base sheet. Lap each ply—4 in.— ....—over preceding ply. Lap ends—6 in.—............ —. After priming deck, embed base sheet in a full hot application of steep asphalt. Broom each ply to assure complete embedment. Mop full width of each lap with steep asphalt.]
   3. At all projections through vapor barrier—and at parapets   *Note 30* —..—, flash up onto projecting element for height of insulation. Flash with— 2—.................—plies of asphalt-saturated fabric, and— 3—.................—courses of plastic roof cement, applied alternately.
   4. ....

F. INSULATION:
   1. Install insulation in accordance with requirements of   *Note 29* manufacturer of—insulation—roofing system—............ ....—.
   2. Insulation shall be—.................in. thick—installed in   *Note 31* ....layer(s)—.

3. Lay no more insulation at one time than can be protected from wetting or other damage by the elements. —Protection of insulation shall be in accordance with requirements of manufacturer of—insulation—roofing system——.......................—.

4. Apply insulation so that longitudinal (continuous) joint is parallel to the short dimension of the roof.

5. Lay insulation with edges in moderate contact, but not forced into place. Stagger end joints.

6. Where more than one layer of insulation is used, stagger joints in each layer, and mop between layers.

[7. Leave—¼ in.—½ in.—.................in.—between insulation and vertical surfaces.]   *Note 32*

8. Install water cutoffs at end of each day's work to protect insulation. Remove water cutoffs before beginning work next day.

9. Protect insulation from damage due to construction operations.

[10. Tape joints between insulation boards as recommended by manufacturer of insulation.]   *Note 33*

11. ....

G. BUILT-UP ROOFING: *Note 29*

1. Install roofing in accordance with roofing manufacturer's specifications—specified hereinbefore—approved by Architect—currently recommended by roofing manufacturer for type of roof deck required in the Contract Documents—and modified herein—.

2. Kettle temperatures for bitumen shall not exceed—475°F for asphalt—ASTM D312, Types II, III, IV 425°F, ASTM D312, Type I—400°F for coal-tar pitch—ASTM D450, Type A..............

3. Apply roofing in shingle fashion so that direction of flow of water is over (and not against) laps.

4. Moppings of bitumen shall be solid under the felts for the full extent required by the roofing specifications, so that at no place in the mopped area shall felt touch felt.

5. Broom—or press—felts into hot bitumen. Lay without wrinkles, buckles,—.............................—or kinks. The finished roofing system shall be free from pockets or blisters.

6. Complete in one operation the application of built-up roofing system—[including—aggregate surfacing—........ ....—]—...—, up to the line of termination at end of day's work.

[7. Repair test spots immediately. —Replace test strip in plastic cement to seal cut surfaces against drippage of bitumen.—Build up roofing over test spot to number of

layers of felts and bitumen required for roof. First ply of
felt in repair shall lap past test spot onto roofing felts—3
in.—.............—all around. Succeeding plies shall lap
preceding ply—3 in.—............................—all around.]

8. ....

H. COMPOSITION BASE FLASHING:
   1. Install composition,—builtup felt—base flashings in
      accordance with requirements of manufacturer of roofing
      system.—Install base flashing where builtup roofing
      abuts vertical surfaces—..
      ....—.

   2. Base flashing shall not extend more than—1 ft 3 in.—
      ....—up vertical surface above level of adjacent
      roofing surface.

   3. ....

I. FINAL SURFACING: *Note 29*
   1. *Mineral Aggregate:*
      a. Install aggregate surfacing in accordance with require-   *Note 35*
         ments of manufacturer of roofing system.
      b. ....
   2. *Smooth Finish:*   *Note 39*
      a. Finish roofing system with a—smooth—glaze-coat—
         ....—surface in accordance with require-
         ments of manufacturer of roofing system.
      b. ....
   3. *Mineral-Surfaced Finish:*
      a. Install mineral-surfaced finish in accordance with
         requirements of manufacturer of roofing system.
      b. ....

J. MINERAL AGGREGATE UNDER—COOLING TOWERS—.........—:   *Note 37*
   1. Install—an additional layer—.......................—of mineral
      aggregate in accordance with requirements of manu-
      facturer of roofing system.
   2. ....

K. ROOF DECKS RECEIVING—PROMENADE TILE DECK   *Note 37*
   COVERING—.......................................—:
   1. Omit mineral aggregate surfacing.
   2. Provide—an additional ply of bituminous-saturated
      roofing felt—..—applied in ac-
      cordance with requirements of manufacturer of roofing
      system.
   3. ....

L. WALKWAYS FOR LIGHT FOOT TRAFFIC ON BUILT-UP
   ROOFING: *Note 29*
   1. Walkways shall consist of—[a continuous strip of............-   *Note 37*
      lb, mineral-surfaced asphalt roofing felt,........in. wide]—
      prefabricated asphalt board—..—.

    [2. Omit roof surfacing material on those portions of built-up roofing to be made into walkways.]

    3. Install walkway surfacing in accordance with requirements of manufacturer of roofing system.

    4. ....

M. ....

05 PROTECTION AND CLEANING:

A. PROTECTION:

    1. Any work or materials damaged during the handling of bitumens and roofing materials shall be restored to perfect condition, or replaced.

    [2. Install protective coverings at all paving and building walls adjacent to hoist and kettles prior to starting any work.]

    [3. Protection for the building wall shall consist of at least—2 full widths of 30-lb (or heavier) roofing felt—6-mil polyethylene—..—lapped not less than—6 in.—................—, and extending from the ground level up to and over the walls and onto the roof deck—..—.]

    [4. Protection shall remain in place for the duration of the roofing work.]

    5. ....

B. CLEANING:

    1. Remove all debris, scraps, containers, and temporary coatings.

    2. Remove bitumens from surfaces other than those requiring bituminous roofing coatings.

    3. ....

0751-06 CLOSEOUT:

A. GUARANTIES AND WARRANTIES: *Note 36*

    [1. *Bond:*

        a. At completion of Project, and prior to final acceptance, Contractor shall furnish to the Architect the roofing manufacturer's ....-year Surety Bond for the builtup roofing—for the composition base flashings used in conjunction with the builtup roofing—for the bituminous connections between the builtup roofing with gravel stops, metal flashings,—................................ ....—..—.]

        b. ....

[2. *Guarantee:*
    a. Contractor shall furnish a written guarantee on the roofing and flashing (both composition and metal) for a—2—................—year period after final acceptance. The guarantee shall provide for repairs of roofing and flashing under this Section where leaking occurs due to faulty materials or workmanship.—The guarantee shall provide for repair or replacement of damage to the building and its finishes resulting from such leaks in the roofing and flashing systems.—The guarantee shall be signed jointly by the Roofing Subcontractor and the General Contractor.—
    b. ....]
3. ....

B. MAINTENANCE AND OPERATING INSTRUCTIONS:
    1. ....
    2. ....

07 SCHEDULES:

08 ALTERNATES: *Note* 13

NOTES FOR SPECIFIC PARAGRAPHS

The following notes apply to paragraphs beside which corresponding *"Note"* designations occur in marginal circles:

Note 1—General and Supplementary Conditions.

Note 2—Article 01 includes several paragraphs which should indicate clearly the work included and the related work not included under this Section. Listing of work to be, and not to be, performed is controversial. These "Scope" paragraphs can be omitted and a brief statement of scope substituted, such as "The work required under this Section consists of Builtup Bituminous Roofing (including Prepared Roll Roofing) and Composition Flashing as shown and scheduled on the drawings and as specified herein."

Note 3—List here work to be furnished and installed under this Section.

Note 4—List here work to be *furnished under this Section* for installation under other Sections of these Specifications or by others not party to this Contract. Indicate who is to install.

Note 5—List here work to be furnished under other Sections of these Specifications or by others not party to this Contract for *installation under this Section.* Indicate who is to furnish. Some work that can fall in this category includes special fasteners for anchoring insulation to steel decks.

Note 6—Listed below are some items of work, related to Built-up Bituminous Roofing which are not included in this work sheet. Should the Project include any of these items, list them under this specification paragraph and designate the Section under which they are covered.   The list follows:

1. Elastomeric roofing
2. Elastomeric flashing
3. Metal cap flashing
4. Through-wall flashing
5. Prefabricated gravel stops
6. Prefabricated reglets
7. Curbs
8. Metal pitch dams
9. Wood nailers at Eaves and Openings in field of insulated roofs, and elsewhere
10. Scuppers
11. Pads for equipment
12. Gutters and downspouts
13. Insulating lightweight aggregate fills
14. Expansion joints

If the Specification Writer elects to include any of the above items in this Section it should be listed under the appropriate paragraph of the "Scope" Article.   The specification text should be amplified accordingly.

Note 7—General Provisions "Notes" should indicate any data on delivery and storage, site and environmental conditions, demolition and salvage, and time or sequence schedules.

Note 8—When bitumens are delivered in tank-car quantities, label requirements would not be applicable.   Instead, determination of compliance would be covered by specification requirements for a certificate of compliance.

Note 9—Prior to final approval of Subcontractors, Suppliers, and Fabricators; lists of all materials proposed by them should be submitted and approved.

Note 10—These letters help to avoid the situations where the Manufacturer is called in too late to inspect the deck and the installation, or when he does not consider the applicator's work generally bondable.

Note 11—Require submission of lists of all proposed systems and materials for approval before approval of Manufacturer or Roofer.

Note 12—List here any tests, certificates, or continuing quality control procedures required.

Note 13—Schedules should be put on the drawings, and a total description of alternates should be put in a part of the Project Manual other than Specifications, since alternates do not become a part of the Contract until made so by the Form of Agreement.   Where it is necessary to refer to schedules or alternates in the technical section of the Specifications, separate Articles should be inserted as shown here.

Note 14—Before listing "Acceptable Manufacturer" verify that each is able to provide the material or system as specified.

Note 15—This form of "Substitutions" paragraph should be used only where a "master substitutions" paragraph is included in the General Conditions, Supplementary Conditions, or Division 1, describing the procedure for submission of requests for material or system substitutions. If data on proposed substitutions is necessary other than the list of materials required, state the additional data required.

Note 16—If more than one set of Built-up Roofing conditions occur on the Project, include list of requirements or description of each.

Note 17—List the characteristics and conditions which are fixed or predetermined for the Project for any reason. For items that are optional to the Roofer and Roofing Manufacturer, note "Optional" or list the options.

Note 18—For specifying minimum material standards permitted in proposed systems and for general use. Include the fixed, predetermined, and optional items only.

Note 19—Cold-applied Roofing Asphalt would be used for prepared Roll Roofing.

Note 20—A heavy asphalt-saturated and coated felt. Base sheet may vary from 40 to 45 lb per 100 sq ft, depending on the manufacturer, and how it is to be used in the roofing system.

Note 21—ASTM D249 covers "Roofing, Asphalt Roll, Surfaced with Mineral Granules." It is made in 3-ft-0-in.-wide rolls. One type has no bare or uncoated edge. Another has a 2-in. bare edge for lapping. Still another has a 4-in. bare edge for lapping. ASTM D371 covers "Roofing Asphalt Roll, Wide Selvage, Surfaced with Mineral Granules." This type is available with 1 ft 5 in. or 1 ft 7 in. of 3-ft-0-in. width of roll coated with mineral granules, and with the remainder of the width bare for nailing and cementing. The necessary quantity and type of nails and lap cement will be furnished if required by the purchaser.

Note 22—This is the "dry sheet" used on wood and other nailable deck to prevent drippage of bitumen and to serve as a base for the roofing membrane.

Note 23—This paragraph provides for specification of a condensed version when it is desired to predetermine only the basic system and the list of acceptable manufacturers. It may be used also to name acceptable standard specification. See Note 28.

Note 24—Over metal decks, roofing manufacturers establish minimum thickness for each type of insulation for various flute opening widths, according to the ability of the insulation to span. See literature of Roofing and Insulation Manufacturers.

Note 25—Choice of fasteners depends on type of deck, insulation, and manufacturers recommendations. Patented self-clinching fasteners, cleats, deformed nails, etc., are available. Type of fasteners may be determined by the approved roofing system. These fasteners may be furnished by the metal deck manufacturer, also. See Note 5.

Note 26—Where more than one Roofing System is used on the Project, use this paragraph to indicate area using particular Roofing System unless it is properly indicated on the drawings.

Note 27—Roofing systems must be adequately secured to substrate to resist uplift by wind. This resistance must be developed by a mechanical bond— either by nailing or by an adhesive bond. Where large cracks occur in the deck or substrate, openings will allow additional air pressure to act on the membrane. Also, the roofing membrane is not a structural element. Adequate, continuous support must be provided for the roofing membrane. Joints in wood sheathing should be tongue-and-groove, shiplap, or splined; and knot holes should be covered with sheet metal. Open joints between precast concrete deck members should be grouted, or stripped with felt. Poured cementitious deck systems should be smooth, patched, and grouted.

Note 28—Where a project can be provided for by specific systems of certain manufacturers, this type of paragraph can be used as a standard of quality.

Note 29—Installation requirements will be furnished with the manufacturer's proposed system. Items are described here to establish a standard of minimum performance.

Note 30—Bituminous vapor barriers are usually 2 layers of 15-lb asphalt-saturated felt or a heavy base sheet bonded to the deck in asphalt bitumen.

Note 31—Thickness of insulation should be shown on the drawings as a part of related details. See discussion of Insulation in Article IIIB3 of these Cover Notes.

Note 32—Some manufacturers do not want a space at the edge of the insulation. Verify with specific manufacturer specified or approved.

Note 33—Some manufacturers of insulation and roofing recommend taping joints of insulation board to provide full bearing for roofing membrane, and to reinforce roofing system at point of stress.

Note 34—Selection of the specific type of bitumen or felt is dependent on the roof slope, type of anchorage of roofing system (nailing or adhesive), insulation, surfacing, climate, conditions of humidity in the building interior, life expectancy of roofing system, cost, maintenance, insurance, and regulating codes.

Weights of felts available are:

- Asphalt-saturated organic: Generally 15 and 30 lb per square.
- Coal-tar-saturated organic: Generally 15 and 30 lb per square.

Glass-fiber fabrics: Available in 6 to 7 lb per square, and 14 to 15 lb per square.

- Asphalt-saturated asbestos felts: 15, 45, and 55 lb per square.
- Bitumen types are as follows:

Asphalt:

Slope per ft, in.	ASTM Designation	Softening point, °F
Up to $\frac{1}{2}$	ASTM D312, Type I	135 to 150
$\frac{1}{2}$ to 3	ASTM D312, Type II	160 to 170
$\frac{1}{2}$ to 6	ASTM D312, Type III	180 to 200
Steeper	ASTM D312, Type IV	205 to 215

Coal-Tar-Pitch:

Slope per ft, in.	ASTM Designation	Softening point, °F
Up to 1	ASTM D450, Type A	140 to 155

Note 35—Specialized glaze-coatings can be specified here.

Note 36—If no bond is required, the type and quality of the roofing system may be specified by a "design life" equivalent to the bond period. Some Specification Writers call for a "bondable" roofing system. A guarantee can be required, with or without a bond.

Note 37—Conditions for providing for more stringent exposure of the roofing system can be listed under 02B3 of the specifications for required submission by the manufacturer of his applicable system for such exposure.

Note 38—To receive some of the fasteners specified in this paragraph, treated wood nailers must be installed at eaves, edges, curbs, walls, and roof openings for securing cants, metal flashing flanges, and backnailed felts or insulation on sloped roofs. This work must be coordinated with the Section on Rough Carpentry.

Note 39—Check with the manufacturer as to where smooth coatings are suitable.

APPROACHES TO FINANCIAL RESPONSIBILITY

There are two basic methods in common use for assigning financial responsibility for roof systems:

- Manufacturer's bond: for 10, 15, or 20 years
- Roofer's guarantee: for 1, 2, or, infrequently, 5 years

These methods may be used singly or in combination. The manufacturer's bond and the roofer's guarantee may be associated. There are also strengthened versions of the above guarantees—for example, when the general contractor and his roofing subcontractor are both required to countersign the manufacturer's bond, a requirement of the State School Building Authority of Georgia.

Another method of seeking good roof performance is to engage a qualified roofing consultant, to advise the architect on the roofing design, and to inspect the roofing contractor's performance on the job.

The Manufacturer's Bond

The most common method of negotiating for financial responsibility is the manufacturer's bond. Historically, the manufacturer's bond appeared in response to a challenge created within the roofing industry. In the early days of built-up roofing, when he doubled as roof applicator, the manufacturer exercised total control over the whole roof construction process—from factory production in the factory to field installation. As the volume of built-up roofing increased during the latter part of the nineteenth century and the early part of the twentieth, manufacturers found their dual role as fabricator-applicator economically impractical. Thus was born the independent roofing contractor, and what was once an integrated responsibility was split in two.

Under the new arrangement, roofing quality frequently declined. Lured by the newly opened opportunities, inexperienced (and often unethical) roofers entered the business and lowered the previous standards set by the manufacturer-applicators. These new roofers sometimes moved from place to place, leaving behind a trail of leaking roofs.

In response to growing roofing troubles, materials manufacturers established standards for manufacturing quality, membrane design, and field practices. In 1905 a pioneering manufacturer originated the specification roof, with its prescription for number of plies, quantity of bitumen, and best application procedures. In 1916 this same manufacturer established a network of approved roofers and, under prescribed conditions, began guaranteeing roofs applied by these roofers. This original roofing bond guaranteed the built-up membrane for 10 years against leaks attributable to material failure or faulty application.

The bonding system began deteriorating after World War II, when the federal government charged that the roofing manufacturers' policies in screening roofers were in restraint of trade. Just when the manufacturers were ordered to ease franchising standards, along came the postwar building boom, with its accelerated construction schedules. Under new competitive pressures, roofing manufacturers competed with one another in licensing roofers, and application standards and manufacturers' inspection services declined.

How the Bond Works   A roofing manufacturer issues the bond, which is backed by a surety company pledged to stand behind the manufacturer's liability for roof failures.   It is designed to assure the owner that:

1. The manufacturer's materials are used.

2. The built-up membrane and the built-up or composition flashing (if covered by supplement to the roofing bond) are installed by a roofer approved by the manufacturer.

3. The manufacturer's representative has inspected the membrane at intervals.

The manufacturer's bond guarantees repairs required only through "ordinary wear and tear of the elements." It is generally voided by "unusual" uses of the roof, e.g., heavy traffic or alterations not authorized by the manufacturer. It is subject to further limitations:

- Insulation, vapor barrier, metal base flashing and metal cap flashing are excluded.

- The bond offers no coverage for damaged structure, finishes, or building contents.

- The liability is limited to $5 to $20 per square at a premium of from $1.50 to $3.00 per square.

The architect should carefully read the roofing bond agreement, noting variations from normal conditions, and explain it to his client (see Fig. 10-1).

Disagreement on bonds   In some aspects the disagreement on the merits of manufacturers' bonds is merely the optimist-pessimist dichotomy: Is the glass half empty or half full? Critics claim that the limited responsibility accepted by the manufacturer delays the repair of roof leaks and complicates the legal proceedings when roof troubles occur; defenders claim that manufacturers' bonds help to establish higher standards of material quality and workmanship.

Eliminating the manufacturer's bond would be disastrous, say its defenders. According to one manufacturer, the death of the manufacturer's bond would result in "absolute chaos"—stultifying technological progress; eradicating standards for materials, application, and workmanship; and destroying the basis for a "sound, profitable, and ethical" conduct of the roofing business.

A Canadian critic of roofing bonds, commenting on his experience in Canada, where roofing bonds were abolished around 1960, concluded, "Instead of the predicted deterioration of standards, there has been considerable improvement in standards at all levels—from design and specifications, through construction. Material standards have been tightened, and there is an increasing demand on the manufacturers by

PHILIP CAREY **PHILIP CAREY CORPORATION**

Know all Men by these presents,

That we, PHILIP CAREY CORPORATION of Cincinnati, Ohio (hereinafter referred to as "Carey"), are held firmly bound* to the owner named below, or the successors and assigns of said owner, in a sum not exceeding the principal amount set forth below, less the aggregate amount expended for repairs made hereunder, for the payment of which we, our successors and assigns hereby bind ourselves by this Bond No._____, said principal amount being $_____.

THE CONDITIONS OF THIS OBLIGATION ARE SUCH that:

Whereas, Carey has sold materials which have been used in constructing a roof described as follows:

Owner: _____

Building location: _____

_____

Building description: _____

No. of squares: _____ Date of completion: _____

Term of Bond from date of completion: Roof_____ years;

Flashing_____ years

Applied by: _____

Address: _____

_____

and

Whereas, subject to the conditions set forth below and elsewhere in this Bond, Carey agrees that during the term of this Bond, it will at its own expense, in the aggregate not exceeding the principal amount hereof, make or cause to be made any repairs that may become necessary solely by reason of ordinary wear and tear by the elements in order to maintain said roof including composition flashing if included above but excluding metalwork and any materials supplied by others, in a watertight condition:

(1) THE OWNER OF SAID ROOF SHALL GIVE CAREY, AT ITS HOME OFFICE, CINCINNATI, OHIO 45215, WRITTEN NOTICE BY REGISTERED MAIL WITHIN SIXTY (60) DAYS AFTER DISCOVERY OF ANY CONDITIONS FOR WHICH CAREY MAY BE RESPONSIBLE UNDER THIS BOND.

(2) Carey shall not be liable in any respect for any damage to the building or to the contents of the building on which said roof is applied, or for any consequential damages.

(3) Carey shall not be liable for any damage to said roof caused by windstorm, lightning, hail or other casualty; or by traffic of any nature over the roof or which results from the use of the roof as a storage area, walking or recreational surface or for any other purpose for which the roof was not designed; or by infiltration, expansion or condensation of moisture in, through or around the walls, copings, building structure or underlying or surrounding materials; or by expansion, contraction or distortion of any metal work or unbonded flashing.

(4) Carey shall not be liable for any damage to said roof or flashing if there is any settlement, cracking, warping, distortion, expansion, contraction, deflection, or other failure of the roof deck, walls or foundation; or if there are any defects in or failure of any materials used as a roof base, or insulation over which said roof is applied; or if there are any defects in design, workmanship or application of materials manufactured by others; or if there is any ponding of water or if underlying materials or structures have failed or ceased to conform to Carey's specifications or other requirements.

(5) All liability of Carey shall cease immediately and automatically if any additions, alterations, or repairs (excluding only emergency temporary repairs) are made except by a Carey approved roofer using Carey materials in accordance with applicable specifications.

(6) Whenever Carey has made repairs called for above costing in the aggregate an amount equal to the principal amount hereof, this bond and Carey's obligations hereunder shall terminate and become void immediately and automatically.

Now Therefore, if Carey, its successors or assigns, shall in all things well and truly perform and observe all and singular the covenants, agreements, stipulations and conditions as shown to be performed and observed by it, then this obligation shall be void; otherwise in full force and effect.

In Witness Thereof, Carey has caused this instrument to be executed by duly authorized representatives this_____ day of _____ 19____

PHILIP CAREY CORPORATION

Countersigned:

_____   _____
Vice President   Attorney-In-Fact

*NOTE: This bond becomes effective only after bills for application and supplies are paid in full to the roofing contractor and material suppliers.

FORM 7118 REV. 7/68   PRINTED IN U.S.A.

FIG. 10-1 *Typical manufacturer's roofing bond.* (*Philip Carey Corp.*)

designers and roofers for even tighter standards more closely related to end use and performance."

Others condemn the elimination of the bond and question the qualifications of roofing consultants in Canada. In effect, roofing contractors are bonding roofs and, through associations, are setting up a fund for repairs, according to these critics. Roofing contractor guarantees in Canada, however, are limited to a maximum of 2 years, with exclusions similar to the manufacturer's bond.

A major charge against roofing bonds is that they lull the architect into a false sense of security, inducing him to relax his specification-writing, design, and inspection responsibilities. There is a widespread misconception that a manufacturer's bond is an insurance policy providing relief in event of a roof leak, regardless of the cause. The surety company does not insure the roof; it cannot be held liable as long as the manufacturer remains in business. It merely guarantees the manufacturer's financial obligation for performing required repairs. Only if the manufacturer goes bankrupt, or for some other reason goes out of business, is the surety company liable for the required repairs.

In rebuttal, manufacturer defenders of the bond maintain that misunderstandings must not be allowed to weigh against the practice of bonding roofs, which, they say, "works to establish higher standards in materials, application, and workmanship."

In general, roofing bonds are more suitable for buildings planned for long-term occupancy. But for large plants where manufacturing improvements and other changes threaten a periodic program of holes cut through the roof, the roofing bond may be unsuitable. A typical manufacturer's bond excludes coverages for damages caused by process discharges or permanent alterations not made by a manufacturer-approved roofer. A steady program of alterations, often performed by the owner's maintenance force, makes it inconvenient to abide by bonding provisions. Such a program may nullify the bond.

Shortening the legal life to 10 years and increasing the liability close to the applied cost of the membrane would bring the roofing bond closer to today's construction realities. Most roofs that perform satisfactorily through the first 2 to 5 years continue to do so for many more years. Thus a shorter term with higher liability would focus the protection when it is most needed. The increased short-term responsibility would inspire the manufacturer to take greater care. The prospect of changing occupancy, which may change the insulation and membrane requirements, enhances the advantages of a short-term bond.

Roofer's Guarantee

The normal 1- or 2-year guarantee may supplement, or, in some cases, replace, a long-term manufacturer's bond. Manufacturers often require a 2-year guarantee from the roofing subcontractor. After an 18-month inspection, defects uncovered by this inspection must be corrected by the roofing subcontractor before the bond takes effect. Failure of the roofing subcontractor in his guarantee to the manufacturer does not relieve the manufacturer of his liability under the bond.

The typical roofer's guarantee requires the roofer to repair leaks resulting "solely from faults or defects in material or workmanship applied by or through the roofer." It excludes the following: all damage attributable to lightning, windstorm, hailstorm, or other unusual phenomena of the elements; foundation settlement; failure or cracking of the roof deck; defects or failure of substrate; vapor condensation under the membrane; faulty construction of parapets, copings, chimneys, skylights, etc.; fire; or clogging of drains. Like the manufacturer's bond, this guarantee excludes liability for damage to building contents or other parts of the structure (see Fig. 10-2).

Occasionally, owners negotiate broader coverage and longer terms in contractors' guarantees—5 years on insulation, vapor barriers, and roof sumps, as well as the roofing membrane.

Although the roofer's guarantee does not carry the psychological weight of the manufacturer's bond, it often proves as effective. Premature roofing failure usually occurs within the 2-year period of the typical guarantee. Moreover, a roofer's guarantee costs less than a manufacturer's bond.

AIA Guaranty for Bituminous Roofing

Still another method for assigning financial responsibility is the "Guaranty for Bituminous Roofing," AIA Document A331. It differs from a manufacturer's bond in requiring a financially unlimited guarantee for repairs. Compared with the ordinary manufacturer's bond, this guarantee gives the owner several advantages, notably:

- Unlimited repairs, covered as required
- Repair of components deteriorating through ordinary wear and tear whether the roof is watertight or not (unlike the roofing bond, which requires evidence of leakage)
- Arrangement for owner's performance of emergency repair without voiding the guarantee
- Reinspection after 2 years

In executing AIA Document A331, the architect must refer to the guarantor manufacturer's specification in the contract documents. The architect's specification should require the roofing subcontractor to inspect surfaces before applying the membrane and to inform the architect of unsatisfactory conditions.

The term of the guarantee should equal the bonding term of the manufacturer's specification.

Because of the provisions committing the manufacturer to financially

Roofing Guarantee

Whereas _____ ,

of _____ ,

herein called "the Contractor," has completed application of the following roof:

Owner: _____

Address of owner: _____

Type and name of building: _____

Location: _____

Area of roof: _____

Date of completion: _____

Date guarantee expires: _____

Whereas, *at the inception of such work the Contractor agreed to guarantee the aforesaid roof against faulty materials or workmanship for a limited period and subject to the conditions herein set forth;*

Now, Therefore, *the Contractor hereby Guarantees, subject to the conditions herein set forth, that during a period of Two (2) years from the date of completion of said roof, it will, at its own cost and expense, make or cause to be made such repairs to said roof and composition flashing resulting solely from faults or defects in materials or workmanship applied by or through the Contractor as may be necessary to maintain said roof in watertight condition.*

This guarantee is made subject to the following conditions:

1. Specifically excluded from this guarantee is any and all damage to said roof, the building or contents caused by lightning, windstorm, hailstorm, or other unusual phenomena of the elements; foundation settlement; failure or cracking of the roof deck; defects or failure of material used as a roof base over which the roof is applied; faulty construction of parapet walls, copings, chimneys, skylights, vents, supports, or other parts of the building; vapor condensation beneath the roof; or fire. If the roof is damaged by reason of any of the foregoing this guarantee shall thereupon become null and void for the balance of the guarantee period unless such damage is repaired by the Contractor at the expense of the party requesting such repairs.

2. The Contractor is not liable for consequential damages to the building or contents resulting from any defects in said roof or composition flashing.

3. No work shall be done on said roof, including, but without limitation, work in connection with flues, vents, drains, sign braces, railings, platforms or other equipment fastened to or set on the roof, and no repairs or alterations shall be made to said roof, unless the Contractor shall be first notified, shall be given the opportunity to make the necessary roofing application recommendations with respect thereto, and such recommendations are complied with. Failure to observe this condition shall render this guarantee null and void. The Contractor shall be paid for time and material expended in making recommendations or repairs occasioned by the work of others on said roof.

4. This guarantee shall become null and void if the roof is used as a promenade or work deck or is sprayed or flooded, unless such use was originally specified and the specification is noted in paragraph 7 below.

5. This guarantee shall not be or become effective unless and until the Contractor has been paid in full for said roof in accordance with the agreement pursuant to which such roof was applied.

6. This guarantee shall become null and void unless the Contractor is promptly notified of any alleged defect in materials or workmanship and provided an opportunity to inspect the roof.

7. Additional conditions or exclusions. _____

_____

_____

In Witness Whereof, this instrument has been duly executed this .... *day of* .... ,

196........

By _____

MEMBER
**MIDWEST
ROOFING
CONTRACTORS
ASSOCIATION**

Approved Guarantee Form No. 1961, Midwest Roofing Contractors Association, Inc.

serving qualified roofing contractors in

ARKANSAS · COLORADO · ILLINOIS · IOWA · KANSAS · MISSOURI · NEBRASKA · OKLAHOMA · SOUTH DAKOTA · TEXAS · WISCONSIN

FIG. 10-2   *Typical roofer's guarantee.   (Midwest Roofing Contractors Association.)*

unlimited repairs far beyond the scope of those guaranteed under the roofing bond, few (if any) manufacturers will sign such guarantees. When available, these guarantees are expensive. It is normally necessary to modify the provisions of AIA Document A331 to get bids on a roofing contract.

Other Approaches to Financial
Responsibility

Several states have tried their own unique methods of assigning financial responsibility for built-up roofs. On much state-administered work, New York State requires the general contractor to furnish a maintenance guarantee, backed by a surety company, for a period not exceeding 10 years. As noted earlier, on state school work Georgia modifies the conventional manufacturer's roofing bond by requiring the general contractor and the roofing subcontractor to become joint principals under the manufacturer's bond.

The Roofing Consultant

The roofing consultant is becoming a more common addition to the architect's team of engineering consultants. Some roofing consultants offer a professional service, advising on design, specifications, and problem investigation. They may also advise architects on roofing system application; and, in conjunction with field inspection, may conduct field and laboratory tests. This kind of consultant is often a professional engineer or architect. Other roofing consultants function more like a conventional testing laboratory—inspecting, reporting, and certifying the roofing subcontractor's workmanship.

TEST CUTS

Test cuts from roofing membranes serve one of two purposes:

▪ A rigorous laboratory diagnosis of an existing roof sample, e.g., a leaking or blistered roof (as described in "Proposed Recommended Practice for Sampling and Examination of Built-up Roofs," ASTM Standards, Part II, p. 896, March, 1968).

▪ On-site examination of a new roofing membrane for conformance with specifications—including visual examination and weighing of specimen to determine bitumen weights (and, in some cases, further laboratory analysis).

Test cuts through new built-up membranes should be made *before* the final pour coat and surfacing are completed—cut through the interply moppings and felts to the substrate (see Fig. 10-3). The difficulty of cutting through the aggregate and the relative ease of checking a pour coat and aggregate after the membrane is completed favors this test procedure over the alternate procedure of cutting through the entire membrane, surfacing and all. Moreover, deficiencies discovered through examination of the test cut are more easily corrected *before* the surfacing is applied.

FIG. 10-3 *Test cut of new roofing, made before surfacing is applied, is inspected for number of plies, adhesive strength of interply bitumen moppings, entrapped moisture, and presence of foreign materials. (Roofing Inspection and Consulting Service.)*

Test cuts are made in 12- × 12-in. squares, or minimum 4- × 36-in. strips. Unlike the square specimen, the long rectangular test cut, taken across the roll direction of the shingled felts, enables the inspector to check the number of plies. Other items checked in specimens cut through new roofing membranes are:

- Character of bitumen, amount and evenness of application
- Qualitative determination of adhesive strength of interply moppings
- Presence of harmful foreign materials
- Evidence of moisture entrapped within the membrane.

The following test-cut procedure, suggested by the national Roofing Contractors Association (1967-R4), is a fairly standard procedure.

Number of test cuts   Two for the first 100 squares or less, plus one for each additional 100 squares or fraction (e.g., five test cuts for a roof area of 350 squares).

Weight tolerance   The test sample shall not weigh less than 15 percent of the correct weight of a 1 sq ft test sample computed from the minimum weights specified for felts in Table 6-2 and from the specified bitumen mopping weights.

Correcting a deficient membrane   If the interply moppings are too light, additional test samples shall be taken to determine the extent of the underweight areas. Deficient areas shall have an additional ply of felt applied in a full mopping of bitumen.

Where free water is discovered between plies, the affected area of built-up membrane shall be removed and rebuilt in dry condition.

The roofing contractor shall make such repairs at no cost to the owner.

Patching cut-out area　　Immediately after its weighing, a *satisfactory* test sample shall be replaced in hot bitumen.　The area shall be covered with three plies of felt, hot mopped in place with the first ply overlapping the cutout area 3 in. on all sides and each succeeding ply overlapping the ply below 3 in. on all sides.

Pour coat and surfacing　　The pour coat and surfacing shall be considered satisfactory if the entire felt surface is covered with the required amount of bitumen and the gravel properly embedded and completely covering the roof surface.

Test cuts are a controversial subject.　Neither roofing contractors nor manufacturers generally advocate the use of test cuts for determining the quality of a new roofing membrane.　Some roofing consultants, on the other hand, not only require test cuts on new membranes, but require that these samples be cut through the surfacing as well as through the felt plies and interply moppings.

ELEVEN

Field Inspection

Inspection is one of the weaker links in the roof construction chain. Many owners are unaware of the need for rigorous field inspection, and some are reluctant to pay the price for such apparently nonproductive work. Good inspection is, however, an essential form of construction insurance, neglected at an owner's peril. One careless or ignorant workman can nullify the most painstaking efforts of the design team. Roofers' guarantees and manufacturers' bond lull many owners into a false sense of security, but because of the difficulty of proving legal responsibility, they are of limited value. Rigorous inspection is the strongest line of defense against poor workmanship.

The obvious basic purpose of inspection is to secure the construction of a good roof. Inspection also has a secondary purpose: to eliminate poor workmanship as a factor in the analysis of any possible subsequent roof problems. Thus the roofer should welcome inspection as a possible means of escaping involvement in possible future litigation. Neither the roofer nor the general contractor, however, should mistakenly assume that inspection in any way relieves them of contractual responsibility to follow the plans and specifications.

For the architect, inspection can be a valuable legal safeguard. If he is sued for faulty design of a failing roof, he must have complete inspection records to avert the risk of taking legal responsibility for poor workmanship or possibly nonconforming materials.

INSPECTION ARRANGEMENTS

There are four basic arrangements for roof inspection:

1. Manufacturer's representative (on roofs covered by a manufacturer's bond)
2. Architect's or owner's representative
3. Roofing consultant
4. Nonprofit industry-sponsored roofing inspection agency

Manufacturer's Representative   Under the terms of most manufacturer's bonds, the manufacturer's representative inspects the installation of flashing and membrane. As part of their bonded roof inspection program, some manufacturers require the inspecting representative to certify that he has made the required inspections—by filing with the manufacturer a certificate signed at the jobsite by the architect, the architect's or owner's representative, or the general contractor's superintendent.

Manufacturer's inspection has a unique drawback. The manufacturer's inspector, often a salesman, may find himself in the awkward spot of rejecting the work of a customer on whose continued good will he depends for future material sales. And some manufacturers have admitted that they are swamped by the sheer volume of roof construction and that at present manpower levels, they cannot do a thorough job of inspecting all their bonded roofs.

Architect or Owner's Representative   Inspection by either the architect's or owner's representative or by a roofing consultant establishes a more objective relation between inspector and roofer. On large projects, a full-time architect's or owner's inspector can inspect roofing application. If not, a roving inspector, retained by the architect, may make periodic site visits. As a jack-of-all-inspection trades, however, the architect's inspector is unlikely to be a master of roofing.

Roofing Consultant   The roofing consultant offers inspection as part of a service that usually includes consultation with the architect on the design of the built-up roofing system. (For more complete description of the roofing consultant's service see Chap. 10, "Specifications and Performance Criteria.")

Inspection Agencies   Nonprofit, industry-sponsored roofing inspection services originated on the West Coast in response to widespread premature roofing failures.   In addition to the jobsite inspection, the service offered by the prototype organization parallels that of the roofing consultant.   It includes preliminary consultation with the architect on the roof-system specification, laboratory-analyzed test cuts, and recommendation on whether the architect should accept or reject the roof after its completion.

PRINCIPLES OF GOOD INSPECTION

The roofing inspector's job is one of the most demanding forms of construction inspection.   The built-up roofing system is the most problem-prone building subsystem manufactured on the jobsite.   The most intricate work on a welded steel truss or a complex set of elevator controls is done in a fabricating shop or factory.   But the roof system is custom-built in the field.   Regardless of the quality of the component materials, the roof's integrity depends on good field workmanship.

The lack of widely accepted standards and test methods handicaps the roofing inspector.   Concrete and welding inspectors, for example, can follow well-established and accepted tolerances, standards, and tests. Determining the bar-replacing accuracy in a flat slab, where depth tolerance is $\frac{1}{4}$ in., is simple and direct.   Inspecting built-up roof construction, however, often requires judgment that cannot be simply resolved with a rule.   Are the fishmouths bad enough to require repair? (See Fig. 11-1.)   Are the felts "wet"?   Is the substrate "smooth"?   Is the aggregate uniformly spread and well embedded in the flood coat?

A good start is crucial.   Firmness at the start establishes the right attitude with one-tenth the effort required to rectify a bad start.   Once

FIG. 11-1   *Fishmouths should be cut and flattened.*

poor workmanship has been tolerated even for a short time, it becomes difficult to correct.

The inspector should obviously know the specifications and details before he sets foot on the roof deck. Nothing will destroy a roofer's respect for an inspector quicker than an early display of naivete or ignorance. The overeager inspector who acts on uncertain knowledge, and then has to back down, has an uphill battle for the rest of that job.

A successful inspection program will generally entail the following:

- A preapplication conference, with inspector (owner's or architect's representative), general contractor, roofer, sheet-metal contractor, and materials manufacturer or agent, held at least 2 days before application begins
- Notification to the inspection agency at least 2 days before application of vapor barrier, insulation, or membrane
- Intensive inspection as each component is installed, especially at the start
- Formal issuance of inspection reports, promptly distributed to all parties concerned, on the progress of the work and its conformance with, or violation of the specifications
- An irregular schedule of inspector appearances
- Formal notice from the contractor when he is ready for final inspection

INSPECTION ALERTS

At the jobsite, the inspector should have access to the following:

- Complete contract documents—specifications and drawings
- Standards referred to in the specifications
- List of approved subcontractors and material suppliers
- Copies of approved shop drawings and submittals
- Copies of correspondence on roof system construction

The inspector's equipment should include a pocket thermometer (scaled from 50 to 500°F), a folding rule, long level, straightedge, and whisk broom.

The inspector should follow a formal routine in making his inspection, using a detailed checklist (see sample checklist, Fig. 11-2).

As roofing materials arrive at the jobsite, the inspector should check labels for conformance with specifications and drawings, approved shop drawings, or change orders.

ROOFING INSPECTION & CONSULTING SERVICE

ROOF INSPECTION REPORT

JOB NAME _____ JOB NO. _____

JOB ADDRESS _____

ROOFING CONTRACTOR _____ DATE _____

INSPECTOR _____ TIME ARRIVED _____ TIME LEFT _____

1. A pre-installation meeting (was) (was not) held prior to start of application.

2. Notice (was) (was not) given that work was in progress today.

3. The Roofing Contractor (was) (was not) working today.

4. Were you able to make an inspection ? (Yes) (No).
   Safe access? (Yes) (No)

5. Were weather conditions O.K. for application of materials?
   (Yes) (No)

6. Was deck condition satisfactory for application of materials?
   (Yes) (No)

7. Could you identify all the materials by manufacturers, labels?
   (Yes) (No)

8. Were materials installed previous day protected from the elements?
   (Yes) (No)

9. Were materials and methods of application same as specified?
   (Yes) (No)

10. If previous work unacceptable, has it been corrected?
    (Yes) (No) (Not applicable)

11. Were test cuts taken this date? (Yes) (No) No. of cuts _____

12. What is estimated area of completed roofing? _____ %

13. Copy of this Report (was) (was not) left at job site _____ With whom? _____

NOTE: EXPLAIN ALL NEGATIVE ANSWERS IN FULL ON BACK OF THIS FORM.

FIG. 11-2   *Formal reports help promote rigorous inspection.   (Roofing Inspection and Consulting Service.)*

Material Storage

1. Prohibit storage of prefabricated insulation or felts directly on the ground or newly poured concrete slabs, where they will absorb moisture. Require pallets over laminated kraft-paper-covered concrete, plywood, or tarpaulin-covered dry earth as a floor for storing insulation or felt rolls.

2. Avoid prolonged site storage.

3. Reject felt rolls compressed into an oval cross section. (They don't lie evenly.) Stack felt rolls on ends.

4. Reject insulation boards with crushed or otherwise damaged edges or corners.

Procedures

1. Inspect roofer's method of operation. Roofing application should begin at far points on deck, working toward areas closest to storage area to minimize traffic over newly applied roof components.

2. Require a tarpaulin covering the wall below the bucket hoist, to prevent staining of the wall with spilled bitumen.

3. Require protection of copings, cant strips, and other building components adjacent to the unloading area.

4. Record unsatisfactory application in precise detail. Note weather conditions, location, and describe violation of specifications or of good roofing practice (see Fig. 11-3).

Structural Deck

1. Check deck for conformance with specified slope, smoothness, dryness, and joint tolerance (see Chap. 3, "Structural Deck"). Make certain that deck is free of dirt and debris. Inspect deck for full compliance *before* roofer starts work (see Fig. 11-4).

2. For steel decks, check deflection, other irregularities in surface. If cold adhesive is specified, check steel especially for transverse dishing, which can prevent bonding with insulation above.

3. For wood decks check application of rosin-sized building paper over deck before hot-mopping, to prevent bitumen drippage.

4. Before hot-mopping poured concrete or gypsum decks, test for sufficient dryness as follows:

Pour a small amount of hot bitumen on deck. If the bitumen, after cooling, can be removed with fingernails, reject this deck for application of the next component—vapor barrier, insulation, or built-up membrane. (The moisture content is too high.) If the cooled bitumen sticks to the deck, and can't be removed by fingers, then the deck is dry enough to receive the mopping.

As a supplementary test, apply hot bitumen as above and observe for frothing or bubbling. If none occurs, the deck is dry enough for installation of the next component.

5. Check elevation of drains, especially for placement too high.

Field Audit*

Construction Consultants, Inc.

Project		Date
Location		Arr: Dep:
Subcontractor	Foreman	Temperature & Weather
Others Contacted:		Bitumen Temperatures C.T.P. STEEP

TODAY'S OPERATIONS (Procedures, Techniques & Equipment in use):   ☐ NO WORK IN PROGRESS   ☐ NO ACCESS TO WORK

. . . .

Substrate	Condition	Preparation
Vapor Barrier	Installation	
Insulation	Installation	
Insulation 2nd Layer	Installation	

MEMBRANE COMPONENTS

Base Ply	Attachment	Installation
Ply Felt/Fabric	Bitumen	Installation
Surfacing	Application	
Flashing, Metal & Accessories		

Check applicable item/s & report in detail on Form 104	104 Field copy given to:
A.☐ Material storage improper. D.☐ Variance in application. B.☐ Material not labeled. E.☐ Work not weathertight. C.☐ Condition detrimental to work. F.☐ Workmanship unsatisfactory	Test samples obtained:
FCC ETG CRW REVIEWED FILE DATE	AUD:

Form 103—(169)   *Data compiled for Construction Consultants, Inc. purposes, use & interpretation only.

FIG. 11-3   *Checklist aids inspector in detecting faulty workmanship.   (Construction Consultants, Inc.)*

6. Check temperature and other weather conditions before application of vapor barrier, insulation, or membrane.   If forecast indicates rain, snow, or high winds, warn general contractor and roofer.

7. Don't permit stacking of materials beyond the design live load on local deck areas.   (Live load generally ranges from 20 to 40 psf.)

FIG. 11-4 *Since reroofing jobs often entail removal of membrane and insulation, great care and rigorous inspection are required to get a clean deck surface.*

Vapor Barriers

1. Do not allow installation of light, flexible plastic-film vapor barrier when high-velocity winds threaten to tear it or prevent smooth adhesion to deck.

2. Make sure that application of vapor barrier (or insulation or membrane) follows soon after quick-setting adhesives are applied.

3. Check spreader-applied ribbons of cold adhesive for proper consistency, thickness, and height.

4. Don't permit application of material with holes or tears.

5. Check side and end laps.

6. Check for perimeter projection where vapor barrier is to be doubled back over insulation.

7. After application, check for punctures, loose seams, or other breaks in the vapor barrier surface.

Thermal Insulation

1. Do not permit installation of wet insulation boards.

2. Limit daily application of insulation to area that can be covered on same day with complete membrane, including top plies and surfacing, or else with a temporary roof scheduled for removal before application of the permanent built-up membrane.

3. If traffic cannot practicably be prohibited from the insulation surface, insist on plywood walkways.

4. See that adjacent units of prefabricated insulation are set with tight joints and that units with broken corners or similar defects are trimmed or discarded (see Figs. 11-5 and 11-6).

5. Check for water cutoffs in insulation boards at the end of day's work. Require removal of vertical face of cutoff detail before work is resumed (see Fig. 11-7).

FIG. 11-5   *All insulation-board joints should be tight, but the continuous longitudinal joints (not the transverse joint shown squeezing up hot bitumen) are normally more critical, and least likely to be tight.   (Acoustical and Insulating Materials Institute.)*

FIG. 11-6   *A broken corner of this insulation board (left) was trimmed and plugged with a triangular fragment cut from scrap (right).   The repair was better than nothing, but the damaged board should not have been installed.   It should have been saved until a roof interruption required a workman to carve a full-sized board for fitting.*

FIG. 11-7   *This mopped-felt cutoff detail has been placed prematurely.   To protect the entire exposed insulation surface from water infiltration, it should have been applied after completion of felt laying over the insulation.   See Fig. 5-6 for correct method of installing cutoff.   (Acoustical and Insulating Materials Institute.)*

6. Check spacing of nails, mechanical anchors, cold adhesives.   On hot-mopped surfaces, check application for conformance with specified spot, channel, or strip mopping, or solid mopping (see Fig. 11-8).

7. Check character of insulation to insure a satisfactory method of bitumen application.

Built-up Membrane

1. For application of built-up membrane, require a smooth, dry, clean substrate, free of projections that might puncture the felts (see Figs. 11-9 and 11-10).   Require sealing of joints in wood and precast decks to prevent bitumen drippage.

2. Require special precautions when built-up membrane is applied at temperatures below freezing.   When ice, frost, or surface moisture is visible on the substrate, prohibit all work until surface has been dried, by natural or artificial means.

3. Check use of heavy mechanical equipment that may puncture the membrane or deflect the deck excessively.

4. Prohibit use of asphalt-saturated felt with coal-tar pitch or vice versa.

5. Require uniform bituminous mopping, without felt touching felt, applied as follows: asphalt—between plies, 20 lb per square; flood coat, 60 lb per square; coal-tar pitch—between plies, 25 lb per square; top

FIG. 11-8   *Sprinkle mopping is a generally unacceptable mode of bonding insulation to any deck, not recommended on the improperly prepared deck surface shown above.*

FIG. 11-9   *Sweeping this wet coated base sheet (note shimmering ladder reflection in water) is not enough to dry it for application of remaining felt plies.   Roof operations should have been (but were not) delayed until the surface had dried.*

coat, 75 lb per square.   Overweight, as well as underweight, moppings are detrimental to the performance and must be avoided.   (For bitumen weight tolerances, see Chap. 6, Field Alert No. 7.)

6. Require pouring, not mopping, of top or flood coat (see Fig. 6-2).

7. Require a visible thermometer, with thermostatic controls, on all kettles, set to the following temperature limits:

Material	Min. temp., °F	Best range, °F	Max. temp., °F	Storage temp., °F
Coal-tar pitch:				
ASTM D450, Type A	300	325–375	400	350
Asphalt:				
ASTM D312, Type I	300	350–400	425	350
ASTM D312, Types II, III, IV	350	400–430	475	425

FIG 11-10 *Suspecting this roof of containing water entrapped betweeen the insulation and felts, the inspector had the felts cut to investigate. The bitumen rolled easily off the glass-fiber insulation, confirming his suspicions. (The bitumen would have adhered tightly to a dry insulation surface.) Subsequently, the roofer removed an 80-sq-ft area of roofing. He repaired it with two coated base sheets, amply overlapped, and three saturated felts, more than equaling the basic three-ply membrane. Where the lower repair felts overlapped the roofing felts, "bucking" water on the slope, an asphalt-saturated fabric tape covered the joint.*

FIG. 11-11 *Manual brooming, close behind the felt-laying machine (which also spreads hot bitumen), smooths the felt, promotes uniform bonding, and helps to expel air or entrapped water vapor through the perforated felt. (The Asphalt Institute)*

FIG. 11-12   *Folded edge of base felt (envelope), nailed to edge strip, prevents bitumen drippage from interply moppings.*

Prohibit use of bitumen heated above the specified maximum.   Require reheating of bitumen too viscous for mopping.   Beware of prolonged storage of bitumen in closed, heated container at temperatures above storage temperatures tabulated above (a practice that can *lower* the softening point of asphalt by 10 to 20°F).

8. Require manual brooming of all felts—saturated and coated— immediately following the mopper or felt-laying machine (see Fig. 11-11).

9. Prohibit phased application with saturated felts, in which one or two plies are laid to protect insulation and top plies placed later.   Final complete surfacing should be placed in the same day's operation as the felts.

10. Limit moisture content for surfacing aggregates as follows:

Crushed stone or gravel, 0.5 percent (by weight)
Roofing slag, 5.0 percent (by weight) in accordance with ASTM
Specification D1863

As a rough field check, aggregate that is wet to the touch should be rejected.   Aggregate must be clean, free of fines, which can destroy its bond to the bitumen.

11. Move dry aggregate to roof at rate of application, stockpiling, if necessary, in small mounds or rows on the finished roofing.   Prohibit storage of aggregate on bare felt.   Do not leave overnight.

12. Require placing of aggregate in *hot* bitumen (vital to good embedment).

13. Require start of roofing application at far points on deck and work toward area where materials are hoisted to roof deck.

14. If test cuts are specified, require that their replacement and patching be in conformance with accepted procedures. (See Chap. 10,

FIG. 11-13 *This wall is defaced by bitumen of low softening point dripping through the edge of the membrane. Drippage can be prevented by folding back and nailing the edge strip (above). A sloped edge also helps to prevent drippage as well as water flow over the eave. (GAF Corporation.)*

FIG. 11-14 *Electrical-resistance heater can prevent cold-process asphalt emulsion from becoming too viscous, for ready application in cold weather.*

FIG. 11-15 *Coated felts for cold-process membrane are laid in place, after precutting and flattening, to get better adhesion than is attainable by unrolling the felts into the bitumen.*

"Specifications and Performance Criteria," for typical test-cut removal and patching procedure.)

15. On slopes of 2 in. per ft or less, require application of felts perpendicular to the slope, starting at the low point.

16. Require backnailing of felts on the following surfaces:

Steep asphalt: slope over $1\frac{1}{2}$ in. per ft
Coal-tar pitch: slope over $\frac{1}{2}$ in. per ft
Dead-level asphalt: slope over $\frac{1}{2}$ in. per ft

17. Require enveloping of perimeter felts to prevent bitumen drippage (see Fig. 11-12 and Fig. 11-13).

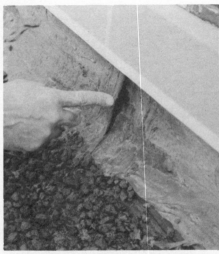

FIG. 11-16   *Faulty flashing installation takes many forms, like the above failure to cement the base flashing to its backing. The resulting horizontal opening exposes this roof-wall intersection to water entry.*

FIG. 11-17   *Flange of pitch-pocket frame should bear on top of the membrane, stripped in with two plies of felt or fabric flashing—not, as shown, on the gravel surfacing.*

Cold-process Membranes

1. Don't permit application of cold-process bitumen at a temperature below the manufacturer's recommendation.

2. When temperature drops below 32°F, require electrical resistance heater or other artificial heating of cold-process bitumen (see Fig. 11-14).

3. Require prior cutting and flattening of saturated and coated felts in cold-process membrane (see Fig. 6-7 and Fig. 11-15).

Flashing

At end of each day's operations, require troweling at top of flashing with flashing cement to close joint and prevent water from entering behind base flashing until counterflashing goes in place (see Figs. 11-16 and 11-17).

TWELVE

New Roofing Membranes

In recent years new roofing membrane materials—plastics or synthetic rubber elastomers applied as fluids or sheets—have begun to challenge the conventional bituminous built-up roofing materials (see Table 12-1).   The trend toward irregular roof surfaces—folded plates, hyperbolic paraboloids, domes, barrel shells—has inspired this development.   The bright, reflective surfaces of some elastomeric materials satisfy architects' esthetic requirements, especially where a curved, sloping surface makes the roof a prominent building feature.   But the new elastomers still constitute only a tiny fraction of the total roofing market.

On an irregular roof surface, these ultralightweight membranes are more easily installed than conventional bituminous materials (see Fig. 12-1).   Scissors, tapes, rollers, brushes, or sprays replace mops, kettles, felt-laying machines, aggregate spreaders, and other more cumbersome equipment used on bituminous built-up roofs.   Flashing details of the new elastomers are also simpler.

The labor-saving potential alone helps to stimulate development of the new elastomers, in the hope that reduced construction labor costs will offset the higher material cost of these new materials.

table 12-1   ELASTOMERIC MEMBRANES

Material		Method of application	Number of coats or sheets
Fluid-applied......	Neoprene-Hypalon	Roller, brush, or spray	2 + 2
	Silicone	Roller, brush, or spray	2
	Polyurethane foam, Hypalon coating	Spray	2
	Clay-type asphalt emulsion reinforced with chopped glass fibers*	Spray	1
Sheet-applied.....	Chlorinated polyethylene on foam	Adhesive	1
	Hypalon† on asbestos felt	Adhesive	1
	Neoprene-Hypalon†	Adhesive	1 + surface paint
	Tedlar‡ on asbestos felt	Adhesive	1
	Butyl rubber	Adhesive	1
Traffic decks......	Silicone plus sand	Trowel	1 + surface coat
	Neoprene with aggregate§	Trowel	1 + surface coat

* Frequently used with coated base sheet.
† Registered trademark of E. I. du Pont de Nemours & Co. for chlorosulfonated polyethylene.
‡ Registered trademark of E. I. du Pont de Nemours & Co. for polyvinyl fluoride.
§ Aggregate may be flint, sand, or crushed walnut shells.

And the hope of avoiding continuing troubles with conventional built-up roofing systems adds still another incentive to the development of new membrane materials.

The new elastomers also offer:

▪ Light weight (normally less than one-tenth the weight of an aggregate-surfaced bituminous built-up membrane)
▪ Adaptability to any roof slope
▪ Good heat reflectivity
▪ High elasticity at moderate temperatures (up to 400 percent elongation strain before breaking)

On the negative side, elastomeric membranes have a more limited range of satisfactory substrate materials than conventional built-up roofing membranes. Many insulation boards are too soft or unstable,

FIG. 12-1   *A sprayed elastomeric covering of white fluid Hypalon coats the concrete shell roof shown above.   (E. I. du Pont de Nemours & Co.)*

and are accordingly banned as substrates for elastomeric membranes by their manufacturers.   Firm deck surfaces are generally the only satisfactory substrates for these new roof coverings.

Some of these new roof coverings installed during the past decade are apparently fulfilling their promise, but limited experience precludes many authoritative recommendations for their design and installation.   For many new elastomers, the manufacturers' recommendations are about the designers' only guide, though some government specifications have been developed.

FLUID-APPLIED ELASTOMERS

Elastomeric membranes applied as fluids solidify through evaporation of volatile solvents and the uniting of the solid residue into a continuous, seamless, weathertight sheet.   After the solvents evaporate, the elastomers resist flame spread, weather deterioration, and most kinds of chemical attack.

Fluid-applied elastomeric membranes are generally better than sheet-applied membranes for roofs of extreme curvature or surfaces penetrated by many vent openings or interrupted with walls, skylights, and mechanical equipment.

The most popular of these fluid-applied membranes combines two synthetic rubbers: *neoprene* (chemically *polychloroprene*) and *Hypalon* (a registered trademark of E. I. du Pont de Nemours & Co., chemically *chlorosulfonated polyethylene*).   Though these materials may be used singly, their complementary qualities are best exploited in combination. For the heavier base coat, which provides waterproofing body, neoprene does the job; for the decorative surface coating, a thinner coating of Hypalon, with its superior color stability and resistance to ozone, sunlight, heat, abrasion, and aging provides a weather-resistant coating.

Fluid neoprene can be brushed, rolled, or sprayed (see Fig. 12-2). A pressure-fed roller is the best application method, and the most economical. A basic neoprene-Hypalon membrane comprises several layers of neoprene covered with one or more coatings of fluid hypalon to a total thickness of about 20 mils (0.020 in.).

These fluid-applied membranes can be applied direct to plywood, to thin-shell concrete and, after suitable joint treatment, to reasonably thick poured concrete slabs. Other structural decks—precast concrete slabs, thin (up to 2½ in. thick) poured slabs on steel joists, metal decks—require fill and leveling to form a suitable surfacing. The best surface for bonding the neoprene primer is a clean, dry, wood-floated concrete surface finished with light steel troweling.

A concrete deck substrate for an elastomeric roof membrane should be water cured, since a sprayed curing agent might prevent bond. The concrete should contain no calcium chloride; subsequent efflorescence might break the bond with the roofing membrane.

The numerous joints in a wood sheathing deck make it unsuitable for a fluid-applied membrane, and the threat of cross-grain shrinkage makes wood sheathing hazardous even for sheet-applied elastomers. The ten-

FIG. 12-2 *Rolling is one of three ways (along with brushing and spraying) to apply fluid neoprene-Hypalon elastomeric roof coverings.* (*E. I. du Pont de Nemours & Co.*)

dency of poured gypsum to crack makes this material a risky substrate for an elastomeric membrane.

Crack control is vital in determining whether a surface is suited to a fluid-applied elastomeric covering. Despite its high elasticity, a neoprene-Hypalon membrane may rupture over hairline cracks. The hazard is especially great in cold weather, when membrane elasticity declines sharply. Thus, if the membrane remains firmly bonded to the deck, a small crack of 0.08 in. opening over an original crack width of 0.008 in. would require a material with an elongation limit of 900 percent to avert rupture. Preliminary laboratory tests on fluid-applied elastomers— neoprene-Hypalon, neoprene epoxy, and acryl styrene resin—indicate that cracks opening 0.04 in. can tear the membrane.

Field performance, however, indicates that these laboratory findings are too severe. Reinforced tapes bridging cracks and joints over $\frac{1}{16}$ in. wide—between precast or prefabricated deck units, metal flashing edges, etc.—can relieve tensile stresses in the membrane. Glass-fiber, nylon, and cotton tape have been used at contraction joints, construction joints, roof ridges, drains, vent stacks, parapets, and other joints requiring flashing.

These tapes have their own special strengths and weaknesses. Because it tends to shrink, cotton tape is not suitable for drainage flashing.

There are two basic neoprene-Hypalon membranes. A general-purpose membrane, suitable for dry-hard substrates, consists of a primer, two coatings of neoprene, topped with two coatings of Hypalon. A heavy-duty version adds a glass-fiber reinforcing mat embedded in the first neoprene coating, and the second coating covers it. A variation on the same theme merely substitutes a continuous glass-fiber roving for the glass-fiber mat.

Ultrathin membranes are suitable for some roofs. One such membrane, designed for thin-shell concrete roofs with no tensile cracking, has two coatings of Hypalon on a neoprene primer.

Silicone rubber, another fluid-applied elastomeric roof covering, retains its flexibility over a wide temperature range—from −85 to 300°F. It also has outstanding durability and resistance to ozone. At least one company markets a silicone covering as a traffic deck.

Silicone elastomers are, however, susceptible to chemical attack from hydrocarbon solvents, acids, strong alkalis, and steam—weaknesses that make them unsuitable for roofs in some industrial atmospheres. Moreover, they attract and hold atmospheric dust, which mars their appearance.

These silicone rubber membranes can be applied to poured or precast concrete decks of any slope or to plywood decks of $\frac{1}{4}$ in. slope or greater. They are not recommended for lightweight concrete fills or gypsum decks.

Moreover, no insulation substrate has yet been found satisfactory for silicone application.

For most membranes, two coats are sufficient: a silicone primer, followed by an undercoat of 10 to 15 mils and a surface coat of 6 to 12 mils, brushed, rolled, or sprayed.   The beveled edges of plywood deck joints are caulked with a silicone sealant applied with a manual caulking gun (see Figs. 12-3 and 12-4).

This same silicone sealant is recommended for minor repairs—cuts, gouges, or improperly sealed joints in the substrate or at roof penetrations.   Such repairs should be made *after* the base coat, but *before* the top coat is applied.

FIG. 12-3   *Sprayed primer prepares caulked plywood roof deck for two-ply, fluid-applied silicone rubber membrane.   (General Electric Co.)*

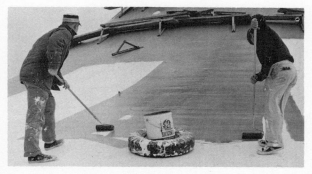

FIG. 12-4   *Surface coating, 6 to 12 mils thick, is rolled onto a slightly thicker base coat for fluid-applied silicone rubber membrane.   (General Electric Co.)*

The sealant can also repair a finished roof with a serious leak or water in the damaged area. But when weather permits, the defective area should be cleaned, primed, and coated with regular silicone rubber for a more visually acceptable repair that will be less dirt retentive than the silicone sealant.

Fluid-applied silicone elastomers are a notable exception to temperature limits for applying most fluid-applied membranes. Silicone rubber can be applied from 0 to 150°F, whereas most fluid-applied coatings are much more restricted. Extreme heat, accelerating the evaporation of solvents, can bubble most fluid-applied coatings; extreme cold delays drying, thus increasing the required waiting time between coatings. Cold also heightens the risk of surface condensation, a threat to the bond between coatings.

SHEET ELASTOMERS

Sheet-applied membranes, the other major category of elastomeric roof coverings, generally cost more than fluid-applied membranes, and they are more difficult to install, especially on irregular surfaces. Sheet application, however, has several important advantages over fluid application. As closely controlled, factory-produced materials, their superior quality often outweighs the advantages of fluid application. Because in many cases it is not continuously bonded to the deck, a sheet can bridge over cracks and thus accommodate greater substrate movement without splitting. Its adhesive can act as a stress transfer medium, absorbing horizontal shear stresses. Moreover, sheets can be applied over a greater range of temperature than is suitable for most fluid applications.

One of the most unusual and successful of these new membranes, sheet neoprene, coated with Hypalon, covers the roof of the Dulles International Airport terminal near Washington, D.C. The neoprene sheet is applied direct to the surface of the precast concrete deck units, which incorporate foamed polystyrene insulation under the deck slabs and between their ribs. To accommodate the membrane stresses resulting from deflections in the long-span inverted concrete arches ($8\frac{1}{2}$-in. maximum deflection at the valley), the architect used a 6-in.-wide neoprene adhesive strip on the concrete surface at the overlapped edges of the sheets. He omitted adhesive from the central portion of the sheet to allow for the large deck movements. The sheets were bonded to each other at end and side laps with the same adhesive.

Following application of the sheet neoprene came the coatings of liquid Hypalon (one aluminum, one gray).

Installed in late 1961, this roof is still performing satisfactorily.

One proprietary laminated sheet covering, a glass-fiber reinforced chlorinated polyethylene sheet backed with open-celled urethane foam, is a true one-ply covering, applied direct to the deck without a base sheet. According to the manufacturer, this covering can be applied to any slope—from 0 to 90 degrees. Its white surface reflects 85 percent of radiant solar heat. The open-celled foamed urethane backing, just over $\frac{1}{16}$ in. thick, can relieve water-vapor pressure by providing horizontal access to edge or stack vents, thus reducing the hazard of blistering.

General recommended application practice for this new one-ply membrane calls for cold-applied field adhesive. For application direct to a steel deck, a cold-applied asphalt emulsion is recommended. The side laps of the 36-in.-wide sheets are 2-in. lap joints (without foam); butted end laps require flashing tape.

Felt-backed elastomeric laminates usually require a coated base sheet as the first ply. One such laminate, a Hypalon sheet backed with asbestos felt, is bonded to the substrate either with a cold adhesive or hot-mopped asphalt. This covering is available in either black or white sheets, whose color can be modified by a fluid application of Hypalon in a broad spectrum of colors (see Fig. 12-5).

Flashing consists of felted or unbacked vinyl sheet.

A similar proprietary product uses Tedlar (another registered du Pont trademark, for polyvinyl fluoride) film for the elastomer with a similar asbestos felt backing, laminated with an elastomeric binder. Over a noncombustible deck this membrane qualifies for a UL Class A flame-spread rating. A coated base sheet is recommended as an underlayment, but is

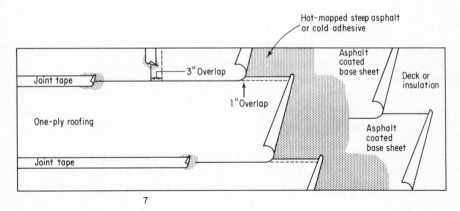

FIG. 12-5 *Laminated elastomeric sheet, faced with 20 mils of white Hypalon and backed with asbestos felt, is bonded to asphalt-coated base sheet with hot-mopped asphalt or cold adhesive.*

not required for most deck surfaces. On a plywood or monolithic concrete deck, a Tedlar-surfaced membrane is generally one-ply. It is bonded to the substrate with either hot-mopped asphalt or a solvent-type, cold asphaltic adhesive.

Side laps of the Tedlar membranes are sealed with a factory-applied lap adhesive, protected with release paper before field use. End laps are made with pressure-sensitive, factory-applied adhesive for the final seal.

Butyl rubber is another kind of elastomeric sheet membrane. It comes in 30-, 45-, and 60-mil thickness, for light, normal, or heavy duty, in widths from 4 to 10 ft or more.

Flashing consists of butyl sheets reinforced with soft copper mesh, easily molded and cut with heavy scissors. Butyl gum tape, 6 in. wide, bonded with a quick-drying adhesive, seals membrane flashing joints. A nondrying, permanently pliable butyl mastic fills occasional voids formed during roof construction.

All elastomeric sheets require more extreme care than conventional built-up roofing felts to prevent puncture by sharp projections in the substrate. Nail heads in wood decks must be flattened and sharp particles removed from concrete surfaces.

With a substrate smooth and dry, the roofer positions and marks the butyl sheets for 6-in. side laps. By folding the sheets back near their longitudinal center line, the roofer bonds them with a central adhesive strip and 6-in. edge-lapped, gum tapes, sandwiched between the lapped sheets. Hand rolling of the joint expels air and ensures good contact between the surfaces.

CAUTIONS WITH ELASTOMERIC MEMBRANES

Moisture penetration (through water-vapor migration, leaks, or wet insulation) is as great a hazard to elastomeric roofing membranes as to conventional bituminous built-up roofs. Under the pressure of evaporating moisture these light, flexible roofs blister more easily than a conventional bituminous built-up roof, and their blisters are more easily punctured. Tending to offset this handicap, repair of the elastomeric coverings is easier.

Cracking in the substrate of an elastomeric membrane is generally more serious than for conventional built-up roofing, at least for fluid-applied coverings (see Fig. 12-6). The drastic reduction of elasticity with falling temperatures in some of these materials (about 800 percent between 72 and 0°F for neoprene-Hypalon) is especially serious.

The sheer novelty of these materials creates problems in field application. A white elastomeric roof designed for spotless architectural appearance requires great care by the applicator. Crews must often be retrained

FIG. 12-6 *Caulking of cracks in concrete shell roof (previously primed for fluid-applied neoprene-Hypalon elastomeric membrane) is essential to prevent membrane splitting. (E. I. du Pont de Nemours & Co.)*

FIG. 12-7 *Sheet-applied elastomeric roofing provides a white reflective roof surface for a St. Louis department store. (GAF Corporation.)*

for applying such roofs. Soft-soled shoes are sometimes worn to reduce the risk of puncturing the freshly coated membrane. Sometimes workmen must wear white sneakers, which must be removed immediately after soiling to avoid tracking smudges over the membrane. Exceptional care is required to prevent the squeezing of adhesives outside a joint tape: dirt adhering to this extruded adhesive can defy most attempts at cleaning it.

Tapes may pose a maintenance problem for sheet-applied elastomeric membranes.

FLUID-APPLIED BITUMINOUS EMULSIONS

On the market some 10 years, a fluid-applied asphalt emulsion reinforced with chopped glass fibers is basically a cold-process bituminous system with some new features that require its inclusion among the new roofing membranes.   A spray gun simultaneously applies the asphalt emulsions and the reinforcing fibers (see Fig. 12-8).

This new fluid-applied system may contain two plies of saturated felt, two plies of asphalt-coated felts, one coated felt and one saturated felt, or, in limited cases, no felt at all.   Membranes often start with application of a solvent-type asphalt primer.   On roofs with sufficient slope and curved surfaces too complex for application of felts, the asphalt emulsion reinforced with woven-glass fibers is sprayed direct to a plywood or concrete deck surface.   The membrane can be finished with a color coating of any shade.

FIG. 12-8   *Triple-barreled spray gun applies chopped glass-fiber reinforcement and asphalt emulsion surfacing over two plies of coated base sheet.   (The Flintkote Co.)*

ALERTS

General

The designer who specifies a new elastomeric membrane should realize that, to some extent, he is entering a new, still experimental field. Extra care is essential when investigating new roofing membranes and inspecting field application.

Design

1. Designers should generally use closed specifications with elastomeric materials. Many problems with fluid-applied elastomeric roofs stem from incorporation of incompatible products manufactured by a competitor of the basic system specified.
2. Carefully investigate details of a roof designed for high visibility. Tapes may discolor in different tones from the membrane material; air pollution may stain the membrane, and dirt on water-repellent surfaces may remain after rain.

Field

Be cautious in specifying temperature and weather conditions for fluid applications. Check manufacturer's recommendations.

Glossary

This glossary contains the technical terms and common expressions of the roofing industry.

ALLIGATORING: Shrinkage cracking of the bituminous surface of built-up roofing, or the exposed surface of smooth-surfaced roofing, in which the loss of volatile oils under solar radiation produces a pattern of deep cracks with the scaly look of an alligator's hide.   It occurs only in unsurfaced bitumen exposed to the weather.

ASPHALT: A dark brown to black, highly viscous, hydrocarbon produced from the residuum left after the distillation of petroleum, used as the waterproofing agent of a built-up roof.   It comes in a wide range of viscosities and softening points—from about 135°F (dead-level asphalt) to 210°F or more (special steep asphalt).   (See Bitumen.)

ASPHALT MASTIC: A mixture of asphaltic material, graded mineral aggregate, and fine mineral matter that can be poured when heated, but requires mechanical manipulation to form.

BACKNAILING: The practice of blind nailing in addition to hot-mopping all the plies to a substrate to prevent slippage on slopes of 1½ in. or more for steep asphalt, ½ in. or more for coal-tar pitch and dead-level asphalt.

BASE SHEET: A saturated or coated felt placed as the first ply in a multi-ply built-up roofing membrane.

BITUMEN: The generic term for an amorphous, semisolid mixture of complex hydrocarbons derived from petroleum or coal.   In the roofing industry there are two basic bitumens: asphalt and coal-tar pitch.   Before application, they are either (1) heated to a liquid state, (2) dissolved in a solvent, or (3) emulsified.

BLISTER: A spongy, raised portion of a roofing membrane, ranging in size from 1 in. in diameter and barely detectable height to as much as 50 sq ft in area and 1 ft high. Blisters result from the pressure of entrapped air or water vapor.

BOND: Adhesive strength preventing delamination of two roofing components. (See also Manufacturer's bond.)

BTU: (British thermal unit) The heat energy required to raise 1 lb of water 1°F in temperature.

BUILT-UP ROOFING MEMBRANE: A continuous, semiflexible roof covering of laminations, or plies, of saturated or coated felts alternated with layers of bitumen, surfaced with mineral aggregate or asphaltic materials.

BUR: Abbreviation sometimes used for built-up roofing membrane.

CANT STRIP: A continuous strip of triangular cross section, fitted into the angle formed by a structural deck and a wall or other vertical surface. The 45-degree slope of the exposed surface of the cant strip provides a gradual transition for base flashing and roofing membrane from a horizontal roof surface to a vertical surface.

CAP FLASHING: (See Flashing.)

CAP SHEET: A mineral-surfaced coated felt (or a coated felt without mineral surfacing) used as the top ply of a built-up roofing membrane.

COAL-TAR PITCH: Dark brown to black solid hydrocarbon obtained from the residuum of the distillation of coke-oven tar, used as the waterproofing agent of dead-level or low-slope built-up roofs. It comes in a narrow range of softening points—from 140 to 155°F.

COATED BASE SHEET (OR FELT): A felt that has previously been "saturated" (impregnated with asphalt) and later coated with harder, more viscous asphalt, which greatly increases its impermeability to moisture.

COLD-PROCESS ROOFING: A bituminous membrane comprising layers of coated felts bonded with cold-applied asphalt roof cement and surfaced with a cutback or emulsified asphalt roof coating.

CONDENSATION: The process through which water vapor (a gas) liquifies as air temperature drops or atmospheric pressure rises. (See Dew point.)

COUNTERFLASHING: (See Flashing.)

COVERAGE: The surface area (in square feet or squares) that should be continuously coated by a specific unit of a roofing material, after allowance is made for a specified lap.

CRACK: A membrane tear produced by bending, often at a wrinkle.

CREEP: (1) Permanent elongation or shrinkage of the membrane resulting from thermal or moisture changes. (2) Permanent deflection of structural framing or structural deck resulting from plastic flow under continued stress or dimensional changes accompanying changing moisture content or temperature.

CUTBACK: An organic, solvent-thinned, soft or fluid cold-process bituminous roof coating or flashing cement.

CUTOFF: A detail designed to prevent lateral water infiltration into the insulation where it terminates at the end of the day's work. A felt strip is (normally) hot-mopped to the stepped contour of the deck, the insulation edge, and the horizontal insulation surface.

DADO: A rectangular groove cut across the grain of a wood blocking member, normally to provide edge canting at the periphery of a roof.

DEAD LEVEL: Absolutely horizontal, of zero slope. (See Slope.)

DELAMINATION: A built-up roofing-membrane failure characterized by separation of the felt plies, sometimes resulting in wrinkling and cracking.

DEW POINT: The temperature at which water vapor starts to condense in cooling air, with no change in atmospheric pressure or vapor content.

EDGE STRIPPING: Application of felt strips cut to narrower widths than the normal 36-in. felt-roll width to start the felt-shingling pattern at a roof edge.

EDGE VENTING: The practice of providing regularly spaced openings at a roof perimeter to relieve the pressure of water vapor entrapped in the insulation.

ELASTOMERIC: Having elastic properties, capable of expanding or contracting with the surfaces to which the material is applied without rupturing.

EMULSION: An intimate mixture of bitumen and water, with uniform dispersion of the bitumen globules achieved through a chemical or clay emulsifying agent.

ENVELOPE: The continuous edge formed by folding an edge base felt over the plies above and securing it to the top felt or, if above-deck insulation is used, to the top surface of the insulation. The envelope thus prevents bitumen drippage through the exposed edge joints of the laminated, built-up roofing membrane and also prevents lateral water infiltration into the insulation.

EQUILIBRIUM MOISTURE CONTENT: The moisture content, expressed as percentage of moisture weight to material weight, at a given temperature and relative humidity.

EXPOSURE: The transverse dimension of a felt not overlapped by an adjacent felt in a built-up roofing membrane. The "exposure" is thus that part of the felt covered directly by the flood coat. The correct felt exposure in a shingled, multi-ply roof is computed by dividing the felt width minus 2 in. by the number of plies (e.g., for two plies of 36-in.-wide felt, the exposure = 36-2/2 = 17 in.).

FALLBACK: A reduction of bitumen softening point, sometimes caused by mixing asphalt with coal-tar pitch or overheating the bitumen. (See Softening-point drift.)

FELT: A fabric manufactured by the interlocking of fibers through a combination of mechanical work, moisture, and heat, without spinning, weaving, or knitting. Roofing felts are manufactured from vegetable fibers (organic felts), asbestos fibers (asbestos felts), or glass fibers (glass-fiber felts).

FISHMOUTH: An opening formed by an edge wrinkle in a felt where it overlaps another felt in a built-up roofing membrane.

FLASHING: Connecting devices that seal membrane joints at expansion joints, drains, gravel stops, and other places where the membrane is interrupted.   *Base flashing* forms the upturned edges of the watertight membrane.   *Cap* or *counterflashing* shields the exposed edges and joints of the base flashing.

FLASHING CEMENT: A trowelable, plastic mixture of bitumen and asbestos (or other inorganic) reinforcing fibers, and a solvent.

FLOOD COAT: The top layer of bitumen in an aggregate-surfaced, built-up roofing membrane.   Correctly applied, it is poured, not mopped, to a weight of 60 lb per square for asphalt, 75 lb per square for coal-tar pitch.

GLAZE COAT: (1) The top layer of asphalt in a smooth-surfaced built-up roof assembly; (2) a thin protective coating of bitumen applied to the lower plies or top ply of a built-up membrane, when the top pouring and aggregate surfacing are delayed.   (See Phased application.)

GRAIN: Weight unit equal to 1/7,000 lb, used in measuring atmospheric moisture content.

GRAVEL: Coarse, granular aggregate, with pieces larger than sand grains, resulting from the natural erosion or crushing of rock.

GRAVEL STOP: Flanged device, normally metallic, designed to prevent loose aggregate from washing off the roof and to provide a finished edge detail for the built-up roofing assembly.

"HOT STUFF" or "HOT":   Roofer's term for hot bitumen.

HYGROSCOPIC: Attracting, absorbing and retaining atmospheric moisture.

HYPALON: A synthetic rubber (chemically chlorosulfonated polyethylene), often used in conjunction with neoprene in elastomeric roof coverings.   (Hypalon is a registered trademark of E. I. du Pont de Nemours & Co.)

INSULATION: See Thermal insulation.

MANUFACTURER'S BOND: A guarantee by a security company that it stands behind a manufacturer's liability to finance membrane repairs occasioned by ordinary wear within a period generally limited to 10, 15, or 20 years.

MEMBRANE: A flexible or semiflexible roof covering, the weather-resistant component of the roofing system.   (See Built-up roofing membrane.)

MINERAL GRANULES: Natural or synthetic aggregate ranging in size from 500 microns to ¼ in. diameter, used to surface cap sheets, slate sheets, and shingles.

MINERAL-SURFACED SHEET: An asphalt-saturated felt, coated on one or both sides and surfaced on the weather-exposed side with mineral granules.

MOPPING: An application of bitumen applied hot with a mop or mechanical applicator to the substrate or to the felts of a built-up roofing membrane.
  Solid mopping: A continuous mopping surface with no unmopped areas.
  Spot mopping: A mopping pattern in which the hot bitumen is applied

in roughly circular areas, generally about 18 in. in diameter, with a grid of unmopped, perpendicular bands.

Strip mopping: A mopping pattern in which the hot bitumen is applied in parallel bands, generally 8 in. wide with 4-in. unmopped spaces.

Sprinkle mopping: A random pattern of heated bitumen beads hurled onto the substrate from a broom or mop.

NEOPRENE: A synthetic rubber (chemically polychloroprene) used in fluid or sheet-applied elastomeric roofing membranes or flashing.

PERLITE: An aggregate used in lightweight insulating concrete and in pre-formed insulating board, formed by heating and expanding silicaceous volcanic glass.

PERM: A unit of water-vapor transmission, defined as 1 grain of water vapor per square foot per hour per inch of mercury pressure difference. (1 in. Hg = 0.491 psi.)

PERMEANCE: An index of a material's resistance to water-vapor transmission. (See Perm.)

PHASED APPLICATION: The practice of applying the felt plies of a built-up roofing membrane in two or more operations, separated by a delay normally of at least 1 day.

PITCH POCKET: A flanged, metal container placed around a column or other roof-penetrating element and filled with bitumen or flashing cement to seal the joint.

PLASTIC CEMENT: See Flashing cement.

PLY: A layer of felt in a built-up roofing membrane: a four-ply membrane has at least four plies of felt at any vertical cross section cut through the membrane. The dimension of the exposed surface ("exposure") of any ply may be computed by dividing the felt width minus 2 in. by the number of plies: thus, the exposed surface of a 36-in.-wide felt in a four-ply membrane should be 8½ in. (See Exposure.)

PRIMER: A thin liquid bituminous solvent applied to a surface to improve the adhesion of heavier applications of bitumen and to absorb dust.

RAKE: The edge of a roof at its intersection with a gable.

REGLET: A groove in a wall or other vertical surface adjoining a roof surface for the embedment of counterflashing.

RELATIVE HUMIDITY: The ratio of the weight (or partial pressure) of water vapor actually diffused through an air-vapor mixture to the saturated weight (or partial pressure) of the water-vapor.

ROLL ROOFING: Coated felts, either smooth or mineral-surfaced.

ROOFER: The roofing subcontractor.

ROOFING SYSTEM: An assembly of interacting roof components designed to weatherproof, and normally to insulate, a building's top surface.

SATURATED FELT: A felt that has been impregnated with bitumen of low softening point—from 100 to 160°F.

SEAL: A narrow counterflashing strip made of bituminous materials.

SELVAGE JOINT: A lapped joint detail for mineral-surfaced cap sheets, in which the mineral surfacing is omitted over the transverse dimension of the overlapping sheet to get better adhesion with the bituminous mopping between the lapped cap sheet surfaces.

SHINGLING: The pattern formed by laying parallel felt rolls with lapped joints so that one longitudinal edge *overlaps* the longitudinal edge of one adjacent felt, whereas the other longitudinal edge *underlaps* the other adjacent felt.   (See Ply.) Shingling is the normal method of applying felts in a built-up roofing membrane.

SLAG: A grayish, porous aggregate left as a residue from blast furnaces and used as surfacing aggregate.

SLIPPAGE: Relative lateral movement of adjacent felt plies in a built-up membrane.   It occurs mainly in sloped roofing membranes, sometimes exposing the lower plies, or even the base sheet, to the weather.

SLOPE: The tangent of the angle between the roof surface and the horizontal, in inches per foot.   The Asphalt Roofing Manufacturers' Association ranks slopes as follows:
   Level: ½-in. maximum
   Low slope: over ½ in. up to 1½ in.
   Steep slope: over 1½ in.

SMOOTH-SURFACED ROOF: A built-up roofing membrane surfaced with a layer of hot-mopped asphalt or cold-applied asphalt-clay emulsion or asphalt cutback, or sometimes with an unmopped, inorganic felt.

SOFTENING POINT: An index of bitumen fluidity.   Asphalt softening point is measured by the "ring-and-ball" test (ASTM D2398).   Coal-tar pitch's softening point is measured by the "cube-in-water" test (ASTM D61).

SOFTENING-POINT DRIFT: A lowering of bitumen softening point, generally caused by prolonged overheating, or mixing asphalt and coal tar pitch.

SOLID MOPPING: (See Mopping.)

SPLIT: A membrane tear resulting from tensile stress.

SPOT MOPPING: (See Mopping.)

SPRINKLE MOPPING: (See Mopping.)

SQUARE: A roof area of 100 sq ft.

STACK VENTING: The practice of providing vertical outlets in the interior areas of a built-up roofing system to relieve the pressure of water vapor entrapped in the insulation.

STRIP MOPPING: (See Mopping.)

STRIPPING: (1) The technique of sealing the joint between metal and built-up membrane with one or two plies of felt or fabric and hot- or cold-applied bitumen. (2) The technique of taping joints between insulation boards.

SUBSTRATE: The surface upon which the roofing membrane is placed—structural deck or insulation.

SYSTEM: (See Roofing system.)

TEDLAR: Polyvinyl fluoride, used as a film surfacing in elastomeric membranes. (Tedlar is a registered trademark of E. I. du Pont de Nemours & Co.)

THERMAL CONDUCTIVITY (k): Heat energy in Btu per hr transferred through a 1-in.-thick 1-sq-ft area of homogeneous material per °F of temperature difference from surface to surface.

THERMAL INSULATION: A material applied to retard the flow of heat through an enclosing surface. For roofs it should have a maximum thermal conductance (C value) of 0.5 (Btu) (hr)/(sq ft) (°F).

THERMAL RESISTANCE (R): An index of a material's resistance to heat transmission, the reciprocal of thermal conductivity k or thermal conductance C.

THERMAL SHOCK: The stress-producing phenomenon resulting from sudden temperature changes in a roof membrane, when, for example, a rain shower follows brilliant sunshine.

VAPOR BARRIER: A material designed to restrict the passage of water vapor through a wall or roof. In the roofing industry, it should be rated at 0.2 perm or less.

VAPOR MIGRATION: The movement of water-vapor molecules from a region of high vapor pressure to a region of lower vapor pressure, penetrating building roofs and walls.

VENT: A stack designed to convey water vapor, or other gas, from inside a building or a building component to the atmosphere.

VERMICULITE: An aggregate used in lightweight insulating concrete, formed by heating and consequent expansion of mica rock.

Appendix

*Included in this Appendix are the introductory
paragraphs to ASTM Standards relevant to the
design of built-up roofing systems.*

Standard Methods of Test for

WATER-VAPOR TRANSMISSION OF MATERIALS IN SHEET FORM

ASTM DESIGNATION: E96-66
ADOPTED, 1966

This Standard of the American Society for Testing and Materials is issued under the fixed designation E96; the number immediately following the designation indicates the year of original adoption or, in the case of revision, the year of last revision. A number in parentheses indicates the year of last reapproval.

SCOPE

1. These methods cover determination of the rate of water-vapor transmission of materials in sheet form. The methods are applicable to materials such as paper, plastic films, and sheet materials in general. The methods are most suitable for specimens $\frac{1}{8}$ in. or less in thickness, but may be used with caution for somewhat thicker specimens. For specimens of materials used in building construction and similar materials greater than $\frac{1}{8}$ in. in thickness, see the Methods of Test for Water Vapor Transmission of Thick

Materials (ASTM Designation C355). Six procedures are provided for the measurement of transmission under different test conditions, as follows:

Procedure A—For use when the materials to be tested are employed in the low range of humidities.

Procedure B—For use when the materials to be tested are employed in the high range of humidities, but will not normally be wetted.

Procedure BW—For use when materials to be tested may in service be wetted on one surface but under conditions where the hydraulic head is relatively unimportant and moisture transfer is governed by capillary and water-vapor diffusion forces.

Procedure C—Conducted at an elevated temperature for use with materials employed in the low range of humidities, and intended to shorten the time of testing of highly impermeable materials. This procedure eliminates, in most cases, the need for refrigeration.

Procedure D—Conducted at an elevated temperature for use with materials employed in the high range of humidities, but not normally wetted, and intended to shorten the time of testing of highly impermeable materials employed in this range. This procedure also eliminates, in most cases, the need for refrigeration.

Procedure E—For use in measuring the WVT at an elevated temperature, with a very low humidity on one side of the sheet and a high humidity on the other side.

Standard Method of Test for

WATER-VAPOR TRANSMISSION OF THICK MATERIALS[1]

ASTM Designation: C355-64

This Standard of the American Society for Testing and Materials is issued under the fixed designation C355; the final number indicates the year of original adoption as standard or, in the case of revision, the year of last revision.

Scope

1. (*a*) These methods cover the determination of water-vapor transmission of materials through which the passage of water vapor may be of importance, such as fiberboards, gypsum, and plaster products, wood products, and plastics. The methods are limited to specimens greater than ⅛ in. (3 mm), and not over 1¼ in. (32 mm) in thickness except as provided in Section 7. Two methods, the desiccant method and the water method, are provided for the measurement of permeance under two different test conditions. Duplication should not be expected between results obtained by the two methods. Results by the water method are frequently much higher. That method should be selected which more nearly approaches the conditions of use.

(*b*) For sheet materials ⅛ in. (3 mm) or less in thickness, see the "Method of Test for Water Vapor Transmission of Materials in Sheet Form" (ASTM Designation E96).[2]

SUMMARY OF METHODS

2. In the desiccant method the test specimen is sealed to the open mouth of a test dish containing a desiccant, and the assembly placed in a controlled atmosphere. Periodic weighings determine the rate of water-vapor movement through the specimen into the desiccant. In the water method, the dish contains pure water, and the weighings determine the rate of vapor movement through the specimen from the water. The vapor pressure difference is nominally the same in both methods.

Standard Specifications for

CORKBOARD THERMAL INSULATION

ASTM DESIGNATION: C352-56
ADOPTED, 1956

This Standard of the American Society for Testing Materials is issued under the fixed designation C352; the final number indicates the year of original adoption as standard or, in the case of revision, the year of last revision.

SCOPE

1. These specifications cover the compositions, sizes, dimensions, and physical properties of baked cork in board form, hereinafter referred to as corkboard, intended for use as thermal insulation on surfaces operating at any temperature below approximately 180°F. For specific applications, the actual temperature limit shall be agreed upon between the manufacturer and the purchaser.

Tentative Specification for

PREFORMED, BLOCK-TYPE CELLULAR POLYSTYRENE THERMAL INSULATION

ASTM DESIGNATION: C578-65T
ISSUED, 1965

This Tentative Specification has been approved by the sponsoring committee and accepted by the Society in accordance with established procedures, for use pending adoption as standard.

Suggestions for revisions should be addressed to the Society at 1916 Race St., Philadelphia, Pa. 19103.

1. SCOPE

1.1 This specification covers the types, grades, physical properties, dimensions, and sizes of cellular polystyrene block or board intended for use as thermal insulation for temperatures up to 170°F (95°C).

1.2 For specific applications, the actual temperature limit shall be agreed upon between the manufacturer and the purchaser.

Tentative Specification for
RIGID PREFORMED CELLULAR URETHANE THERMAL INSULATION

ASTM DESIGNATION: C591-66T
ISSUED, 1966

This Tentative Specification has been approved by the sponsoring committee and accepted by the Society in accordance with established procedures, for use pending adoption as standard. Suggestions for revisions should be addressed to the Society at 1916 Race St., Philadelphia, Pa., 19103.

1. SCOPE

1.1 This specification covers the composition, properties, standard sizes, and dimensional tolerances of preformed rigid cellular urethane thermal insulation intended for use on pipes and flat surfaces operating within the temperature range of −100°F (−73.3°C) to +230°F (110°C). For questionable applications, the actual temperature limits shall be agreed upon between the manufacturer and the purchaser. Only production urethane block and pipe insulation are covered by this specification.

Note 1: The application of a vapor barrier is recommended in conjunction with this insulation where service temperatures are generally below ambient.

Note 2: For cellular urethane other than thermal insulation, see ASTM Specification D2341, for Rigid Urethane Foam.

Standard Specification for
CELLULAR GLASS BLOCK AND PIPE THERMAL INSULATION

ASTM DESIGNATION: C552-66
ADOPTED, 1966

This Standard of the American Society for Testing and Materials is issued under the fixed designation C552; the number immediately following the designation indicates the year of original adoption or, in the case of revision, the year of last revision. A number in parentheses indicates the year of last reapproval.

1. SCOPE

1.1 This specification covers the composition, sizes, dimensions, and physical properties of cellular glass block and pipe thermal insulation intended for use on surfaces operating at temperatures between −300 and +800°F (−184 and +427°C).

1.2 For specific applications, the actual temperature limit shall be agreed upon between the manufacturer and the purchaser.

Standard Methods of
TESTING CELLULAR GLASS INSULATING BLOCK

ASTM DESIGNATION: C240-61
ADOPTED, 1956; REVISED, 1961

This Standard of the American Society for Testing Materials is issued under the fixed designation C240; the final number indicates the year of original adoption as standard or, in the case of revision, the year of last revision.

SCOPE

1. These methods cover the testing of cellular glass insulating block for density, water absorption, compressive strength, flexural strength, and thermal conductivity.

Tentative Specification for
MINERAL-FIBER BLOCK AND BOARD THERMAL INSULATION

ASTM DESIGNATION: C612-67T

This Tentative Specification has been approved by the sponsoring committee and accepted by the Society in accordance with established procedures, for use pending adoption as standard. Suggestions for revisions should be addressed to the Society at 1916 Race St., Philadelphia, Pa. 19103.

1. SCOPE

1.1 This specification covers the determination of composition, physical properties, and dimensions of mineral fiber (rock, slag, or glass) block and board intended for use as thermal insulations on surfaces at temperatures below ambient and above ambient up to 1800°F (982°C). For specific applications the actual temperature limit shall be agreed upon between the manufacturers and the purchaser.

Note 1: The values stated in U.S. customary units are to be regarded as the standard. The metric equivalents of U.S. customary units given in the body of the standard may be approximate.

Note 2: For applications below ambient, the installation of a vapor barrier is recommended.

Standard Specifications for

STRUCTURAL INSULATING BOARD MADE FROM VEGETABLE FIBERS

ASTM DESIGNATION: C208-60
ADOPTED, 1960

This Standard of the American Society for Testing Materials is issued under the fixed designation C208; the final number indicates the year of original adoption as standard or, in the case of revision, the year of last revision.

SCOPE

1. These specifications cover structural insulating board used principally in building construction. Seven classes of board are covered, as follows:
Class A—Building board.
Class B—Lath (for plaster base).
Class C—Roof insulation board.
Classes D and *F*—Interior boards.
  Classes *D* (1) and *F* (1)—Interior finish board.
  Classes *D* (2) and *F* (2)—Panels (or tileboard).
  Classes *D* (3) and *F* (3)—Plank.
Class E—Sheathing.
Class G—Shingle backer.

Standard Specifications for

STRUCTURAL INSULATING FORMBOARD MADE FROM VEGETABLE FIBERS

ASTM DESIGNATION: C532-66
ADOPTED, 1966

This Standard of the American Society for Testing and Materials is issued under the fixed designation C532; the number immediately following the designation indicates the year of original adoption or, in the case of revision, the year of last revision. A number in parentheses indicates the year of last reapproval.

SCOPE

1. This specification covers structural insulating formboard for use as a permanent form for poured-in-place roof construction of reinforced gypsum

or lightweight insulating concretes. Insulating formboard is a fabricated structural insulating board used primarily on industrial buildings, warehouses, schools, and commercial buildings, and is usually supplied in nominal 1-in. ($\pm \frac{5}{64}$-in.) thickness and 32-in. width, with lengths from 4 to 12 ft. It is also fabricated in thicknesses, widths, and lengths for specific job conditions. The interior surface may be plain or factory finished. Insulating formboard is placed on roof framing members, usually subpurlins, and the roof deck material is poured in place. Insulating formboard is used as a permanent form that remains as an integral part of a poured-in-place roof deck. This specification covers only nominal 1-in.-thick plain or factory-finished materials.

Standard Methods of

TESTING STRUCTURAL INSULATING BOARD MADE FROM VEGETABLE FIBERS

ASTM Designation: C209-60(1966)
ADOPTED, 1960; REAPPROVED, 1966

This Standard of the American Society for Testing and Materials is issued under the fixed designation C209; the final number indicates the year of original adoption as standard or, in the case of revision, the year of last revision.

Scope

1. These methods cover procedures for testing the following properties of structural insulating board made principally from wood, cane, or other vegetable fibers. The methods appear in the following order:

	Sections
Thickness	5 to 8
Thermal conductivity	9
Transverse strength	10 to 13
Deflection at specified minimum load	14 and 15
Deflection at breaking load	16 and 17
Tensile strength parallel to surface	18 to 21
Tensile strength perpendicular to surface	22 to 25
Water absorption	26 to 29
Linear expansion	30
Vapor transmission	31
Inclined panel flame test	32 to 36
Direct fastener withdrawal test	37 to 39
Lateral fastener resistance test	40 to 42
Direct fastener pull-through resistance test	43 to 45
Moisture content and density	46 to 48
Racking load	49
Concentrated load strength	50 to 53

Standard Method of Test for

STRUCTURAL INSULATING ROOF DECK

ASTM Designation: D2164-65
Adopted, 1965

This Standard of the American Society for Testing and Materials is issued under the fixed designation D2164; the final number indicates the year of original adoption as standard or, in the case of revision, the year of last revision.

SCOPE

1. These methods cover procedures for determining the following properties of structural insulating roof deck. In all structural tests the specimens are loaded as beams with the finished (ceiling) face in tension.

	Sections
Equivalent uniform load	6 to 10
Concentrated load	11 to 14
Sustained uniform load (sag)	15 to 18
Impact load	19 to 22
Resistance to cyclic exposure	23 to 27
Thermal conductance (method C177)	28 to 30
Vapor permeance (methods C355)	31 to 35

Standard Specification for

VERMICULITE LOOSE-FILL INSULATION

ASTM Designation: C516-67

This Standard of the American Society for Testing and Materials is issued under the fixed designation C516; the number immediately following the designation indicates the year of original adoption or, in the case of revision, the year of last revision. A number in parentheses indicates the year of last reapproval.

SCOPE

1. This specification covers the composition and physical properties of expanded or exfoliated vermiculite loose-fill insulation, when used in the temperature range of −50 to 1400°F, and also includes the testing procedures by which the acceptability of the material may be determined. It is not the intent of this specification to limit the use of this material to this temperature range.

Standard Specification for

PERLITE LOOSE-FILL INSULATION

ASTM DESIGNATION: C549-67

This Standard of the American Society for Testing and Materials is issued under the fixed designation C549; the number immediately following the designation indicates the year of original adoption or, in the case of revision, the year of last revision. A number in parentheses indicates the year of last reapproval.

1. SCOPE

1.1 This specification covers the composition and physical properties of expanded perlite loose fill insulation when used in the temperature range of -50 to $1400°F$. It also includes the testing procedures by which the acceptability of the material may be determined. It is not the intent of this specification to limit the use of this material to this temperature range.

Standard Specifications for

ASPHALT FOR USE IN CONSTRUCTING BUILT-UP ROOF COVERINGS

ASTM DESIGNATION: D312-64(1958)
ADOPTED, 1941; LAST REVISED, 1964; REAPPROVED, 1958

This Standard of the American Society for Testing and Materials is issued under the fixed designation D312; the final number indicates the year of original adoption as standard or, in the case of revision, the year of last revision.

SCOPE

1. These specifications cover asphalt intended for use as hot-cement and mopping coat in the construction of built-up roof coverings for roofs surfaced in various manners, laid either over boards or concrete on various inclines.

Standard Method of Test for

SOFTENING POINT OF ASPHALT (BITUMEN) AND TAR IN ETHYLENE GLYCOL (RING AND BALL)

ASTM DESIGNATION: D2398-68

This Standard of the American Society for Testing and Materials is issued under the fixed designation D2398; the number immediately following the designation indicates the year of original adoption or, in the case of revision, the year of last revision. A number in parentheses indicates the year of last reapproval.

1. SCOPE

1.1 This method covers the determination of the softening point of asphalt (bitumen) and tar, including tar pitches, in the range of 85 to 347°F (30 to 175°C) using the ring-and-ball apparatus in an ethylene glycol bath.

1.2 Results obtained by this method differ from those obtained by ASTM Method D36 and IP Method 58, Test for Softening Point of Asphalts and Tar Pitches (ring-and-ball apparatus), by the following amounts:

1.2.1 Below 176°F (80°C) results from this method will be 4.5°F (2.5°C) higher than Method D36, or 8.5°F (4.0°C) higher than IP58.

1.2.2 Above 176°F (80°C) results from this method will be 2.5°F (1.5°C) lower than Method D36, but will be similar to those of IP58.

2. SUMMARY OF METHOD

2.1 A steel ball of specified weight is placed upon a disk of sample contained within a horizontal, shouldered, metal ring of specified dimensions. The assembly is heated in an ethylene glycol bath at a uniform, prescribed rate and the softening point taken as the temperature at which the sample becomes soft enough to allow the ball, enveloped in the sample material, to fall a distance of 1 in. (25.4 mm).

3. SIGNIFICANCE

3.1 Asphalt (bitumen) and tar, including tar pitches, do not change from the solid state to the liquid state at any definite temperature, but gradually become softer and less viscous as the temperature rises. For this reason, the determination of softening point must be made by a fixed, arbitrary, and closely defined method if the results obtained are to be comparable.

Standard Specifications for

COAL-TAR PITCH FOR ROOFING, DAMPPROOFING, AND WATERPROOFING

ASTM DESIGNATION: D450-41
REAPPROVED 1965

This Standard of the American Society for Testing and Materials is issued under the fixed designation D450; the number immediately following the designation indicates the year of original adoption or, in the case of revision, the year of last revision. A number in parentheses indicates the year of last reapproval.

SCOPE

1. These specifications cover coal-tar pitch suitable for use as a mopping coat in the construction of built-up roofs surfaced with slag or gravel, as a mopping coat in dampproofing, or as a plying or mopping cement in the construction of a membrane system of waterproofing.

Standard Methods of Test for

SOFTENING POINT OF PITCH (CUBE-IN-WATER METHOD)

ASTM DESIGNATION: D61-68

This Standard of the American Society for Testing and Materials is issued under the fixed designation D61; the number immediately following the designation indicates the year of original adoption or, in the case of revision, the year of last revision. A number in parentheses indicates the year of last reapproval.

1. SCOPE

1.1 This method covers the determination of the softening point of pitches that have softening points below 80°C by this method. Pitches of higher softening point are tested by ASTM Method D2319, Test for Softening Point of Pitch, Cube-in-Air Method.

Standard Specifications for

CREOSOTE FOR PRIMING COAT WITH COAL-TAR PITCH IN DAMPROOFING AND WATERPROOFING

ASTM DESIGNATION: D43-41
REAPPROVED, 1965

This Standard of the American Society for Testing and Materials is issued under the fixed designation D43; the number immediately following the designation indicates the year of original adoption or, in the case of revision, the year of last revision. A number in parentheses indicates the year of last reapproval.

SCOPE

1. These specifications cover creosote primer for use, when specified, with coal-tar pitch in dampproofing and waterproofing below or above ground level, for application to concrete and masonry surfaces.

Standard Specifications for

COAL-TAR SATURATED ROOFING FELT FOR USE IN WATERPROOFING AND IN CONSTRUCTING BUILT-UP ROOFS

ASTM DESIGNATION: D227-56(1965)

ADOPTED, 1927; LAST REVISED, 1956; REAPPROVED, 1965

This Standard of the American Society for Testing and Materials is issued under the fixed designation D227; the final number indicates the year of original adoption as standard or, in the case of revision, the year of last revision.

SCOPE

1. These specifications cover coal-tar saturated felt, either 36 or 32 in. in width, composed of roofing felt saturated, but not coated, with refined coal tar, for use in the membrane system of waterproofing and in the construction of built-up roofs.

Standard Specifications for

ASPHALT-SATURATED ROOFING FELT FOR USE IN WATERPROOFING AND IN CONSTRUCTING BUILT-UP ROOFS

ASTM DESIGNATION: D226-68

This Standard of the American Society for Testing and Materials is issued under the fixed designation D226; the final number indicates the year of original adoption as standard or, in the case of revision, the year of last revision.

SCOPE

1. These specifications cover asphalt-saturated felts, either with or without perforations, 36 or 32 in. in width, composed of roofing felt saturated, but not coated, with asphalt for use in the membrane system of waterproofing and in the construction of built-up roofs.

TYPES

2. Asphalt-saturated felts covered by these specifications are of two types: namely,

15-lb type
30-lb type

Tentative Specification for

ASPHALT-IMPREGNATED GLASS FIBER MAT (FELT)

ASTM DESIGNATION: D2178-63T
ISSUED, 1963

This Tentative Specification has been approved by the sponsoring committee and accepted by the Society in accordance with established procedures, for use pending adoption as standard. Suggestions for revisions should be addressed to the Society at 1916 Race St., Philadelphia, Pa. 19103.

SCOPE

1. This specification covers glass-fiber mats impregnated to varying degrees with asphalt for use in the construction of built-up roofs and in the membrane system of waterproofing (see note).

Note: These asphalt-treated glass-fiber mats are frequently referred to in the industry as "glass-fiber felts" and the methods of test used for determining their properties shall be the same, except where otherwise specified herein, as those prescribed in the "Methods of Sampling and Testing Felted and Woven Fabrics with Bituminous Substances for Use in Waterproofing and Roofing" (ASTM Designation D146).

TYPES

2. Asphalt-treated glass-fiber mats covered by this specification are of two types:

Type I, 8-lb type
Type II, 15-lb type

Standard Specifications for

ASPHALT-SATURATED ASBESTOS FELTS FOR USE IN WATERPROOFING AND IN CONSTRUCTING BUILT-UP ROOFS

ASTM DESIGNATION: D250-68

This Standard of the American Society for Testing and Materials is issued under the fixed designation D250; the final number indicates the year of original adoption as standard or, in the case of revision, the year of last revision.

SCOPE

1. These specifications cover asphalt-saturated asbestos felts, either with or without perforations, 36 or 32 in. in width, composed of asbestos felt, saturated, but not coated, with asphalt for use in the membrane system of waterproofing and in the construction of built-up roofs.

Standard Specifications for

ASPHALT-SATURATED AND COATED ASBESTOS FELTS FOR USE IN CONSTRUCTING BUILT-UP ROOFS

ASTM DESIGNATION: D655-47(1965)
ADOPTED, 1947; REAPPROVED, 1965

This Standard of the American Society for Testing and Materials is issued under the fixed designation D655; the final number indicates the year of original adoption as standard or, in the case of revision, the year of last revision.

SCOPE

1. These specifications cover asphalt-saturated and asphalt-coated asbestos felts in sheet form for use in the construction of built-up roofs. They may be either 36 or 32 in. in width and shall be composed of asbestos roofing felt saturated and coated on one or both sides with asphalt. The coated sides shall be surfaced with mineral matter.

TYPES

2. Asphalt-saturated and asphalt-coated asbestos felts covered by these specifications are of two types; namely,

> 20-lb type
> 50-lb type

Tentative Specification for

ASPHALT-BASE SHEET FOR USE IN CONSTRUCTION OF BUILT-UP ROOFS

ASTM DESIGNATION: D2626-67T

This Tentative Specification has been approved by the sponsoring committee and accepted by the Society in accordance with established procedures, for use pending adoption as standard.

Suggestions for revisions should be addressed to the Society at 1916 Race St., Philadelphia, Pa., 19103.

1. SCOPE

1.1 This specification covers Type I and II asphalt-base sheet 36 in. (91.4 cm) in width composed of roofing felt saturated and coated on both sides with asphalt and surfaced on the top side with fine mineral surfacing for use as the first ply of a built-up roof, or as a vapor barrier under roof insulation.

Standard Specifications for

ASPHALT ROLL ROOFING SURFACED WITH MINERAL GRANULES

ASTM DESIGNATION: D249-68

This Standard of the American Society for Testing and Materials is issued under the fixed designation D249; the final number indicates the year of original adoption as standard or, in the case of revision, the year of last revision.

SCOPE

1. These specifications cover asphalt roofing in sheet form, 36 in. in width, composed of roofing felt saturated and coated on both sides with asphalt, and surfaced on the weather side with mineral granules.

Standard Specifications for

ASPHALT ROLL ROOFING SURFACED WITH POWDERED TALC OR MICA

ASTM DESIGNATION: D224-68

This Standard of the American Society for Testing and Materials is issued under the fixed designation D224; the final number indicates the year of original adoption as standard or, in the case of revision, the year of last revision.

SCOPE

1. These specifications cover asphalt roofing in sheet form, 36 in. in width, composed of roofing felt saturated and coated on both sides with asphalt and surfaced on the weather side with powdered mineral matter such as talc or mica.

Standard Specifications for

WIDE-SELVAGE ASPHALT ROLL ROOFING SURFACED WITH MINERAL GRANULES

ASTM DESIGNATION: D371-58
ADOPTED, 1958
REAPPROVED IN 1965 WITHOUT CHANGE

This Standard of the American Society for Testing Materials is issued under the fixed designation D371; the final number indicates the year of original adoption as standard or, in the case of revision, the year of last revision.

SCOPE

1. These specifications cover asphalt roofing in sheet form, 36 in. in width, composed of asphalt-saturated roofing felt coated on approximately one half of the width of the weather side with asphalt, and surfaced on the coated portion with mineral granules. This roofing is used as a cap sheet in the construction of built-up roofs.

Standard Specifications for

WOVEN GLASS FABRICS TREATED WITH BITUMINOUS SUBSTANCES FOR USE IN WATERPROOFING

ASTM DESIGNATION: D1668-63
ADOPTED, 1963

This Standard of the American Society for Testing and Materials is issued under the fixed designation D1668; the final number indicates the year of original adoption as standard or, in the case of revision, the year of last revision.

SCOPE

1. These specifications cover bituminized glass fabric, composed of woven glass cloth treated with either asphalt or coal-tar pitch, as specified by the purchaser, for use in the membrane system of waterproofing.

Standard Methods of

SAMPLING AND TESTING FELTED AND WOVEN FABRICS SATURATED WITH BITUMINOUS SUBSTANCES FOR USE IN WATERPROOFING AND ROOFING

ASTM DESIGNATION: D146-65
ADOPTED, 1959; REVISED, 1965

This Standard of the American Society for Testing and Materials is issued under the fixed designation D146; the final number indicates the year of original adoption as standard or, in the case of revision, the year of last revision.

SCOPE

1. These methods cover the sampling and examination of felts or woven fabrics, saturated but not coated with asphalt or coal-tar materials, for use in the membrane system of waterproofing or for the construction of built-up roof coverings.

Standard Specifications for

WOVEN COTTON FABRICS SATURATED WITH BITUMINOUS SUBSTANCES FOR USE IN WATERPROOFING

ASTM DESIGNATION: D173-68

This Standard of the American Society for Testing and Materials is issued under the fixed designation D173; the final number indicates the year of original adoption as standard or, in the case of revision, the year of last revision.

SCOPE

1. These specifications cover bituminized cotton fabric, composed of woven cotton cloth waterproofed with either asphalt or coal-tar pitch, as specified by the purchaser, for use in the membrane system of waterproofing.

Standard Specifications for

ASPHALT-BASE EMULSIONS FOR USE AS PROTECTIVE COATINGS FOR BUILT-UP ROOFS

ASTM DESIGNATION: D1227-65

This Standard of the American Society for Testing and Materials is issued under the fixed designation D1227; the final number indicates the year of original adoption as standard or, in the case of revision, the year of last revision.

SCOPE

1. These specifications cover asphalt-base emulsions capable of being spray- or brush-applied in relatively thick films as a protective coating for roof surfaces having inclines of not less than ½ in. per ft (13 mm per 305 mm).

TYPES

2. *Type I.*—Type I includes asphalt-base emulsions prepared with mineral colloid emulsifying agents.

Type II.—Type II includes asphalt-base emulsions prepared with chemical emulsifying agents.

Standard Methods of

TESTING ASPHALT-BASE EMULSIONS FOR USE AS PROTECTIVE COATINGS FOR BUILT-UP ROOFS

ASTM DESIGNATION: D1167-65

This Standard of the American Society for Testing and Materials is issued under the fixed designation D1167; the final number indicates the year of original adoption as standard or, in the case of revision, the year of last revision.

SCOPE

1. (*a*) These methods are intended for the examination of asphalt emulsions for use as protective coatings for built-up roofs, composed principally of an asphaltic base, water, and an emulsifying agent.

(*b*) These methods cover the following tests:

Standard Specifications for

MINERAL AGGREGATE FOR USE ON BUILT-UP ROOFS

ASTM DESIGNATION: D1863-64

This Standard of the American Society for Testing and Materials is issued under the fixed designation D1863; the final number indicates the year of original adoption as standard or, in the case of revision, the year of last revision.

SCOPE

1. These specifications cover the quality and grading of crushed stone, crushed slag, and water-worn gravel suitable for use as coarse mineral aggregate on built-up roofs.

Standard Methods of

FIRE TESTS OF ROOF COVERINGS

ASTM DESIGNATION: E108-58
ADOPTED, 1958

This Standard of the American Society for Testing Materials is issued under the fixed designation E108; the final number indicates the year of original adoption as standard or, in the case of revision, the year of last revision.

SCOPE

1. (*a*) These methods are intended to measure the fire-retardant characteristics of roof coverings against fire originating outside the building on which they have been installed. They are applicable to roof coverings intended for installation on either combustible or noncombustible decks, and when applied in a manner to prevent leakage.

(*b*) Three methods of test are included as follows:

Method A: Intermittent flame exposure test,

Method B: Spread of flame test

Method C: Burning brand test.

Bibliography

NOTE TO READER: *The following references were among those consulted by the author. The reader will, however, find many contradictions between this manual and the referenced works and among different referenced works.*

Books

1. *ASHRAE Handbook of Fundamentals*, prepared by the American Society of Heating, Refrigerating and Air Conditioning Engineers, Inc., 345 E. 47th Street, New York 10017, 1967.
2. *Handbook of Airconditioning, Heating and Ventilating*, 2d ed., by Clifford Strock and Richard L. Koral (eds.), The Industrial Press, New York, 1965.
3. *Engineering Properties of Roofing Systems*, Symposium of 69th Annual Meeting, American Society of Testing and Materials, Atlantic City, June 26 to July 1, 1966.
4. *Handbook of Air Conditioning System Design*, prepared by the Carrier Air Conditioning Co., McGraw-Hill Book Company, New York, 1965.
5. *Mechanical and Electrical Equipment for Buildings*, by William J. McGuinness and Benjamin Stein, John Wiley & Sons, Inc., New York, 1964.
6. *Principles for Air Conditioning Practice*, by W. F. Stoecker, The Industrial Press, New York, 1968.
7. *Engineering Manual*, 2d ed., by Robert H. Perry (ed.), McGraw-Hill Book Company, New York, 1967.
8. *Building Construction Handbook*, 2d ed., by Frederick S. Merritt, McGraw-Hill Book Company, New York, 1965.
9. *Rigid Plastics Foams*, 2d ed., by T. H. Ferrigno, Reinhold Publishing Corporation, New York, 1967.
10. *Plastics in Building*, by Irving Skeist (ed.), Reinhold Publishing Corporation, New York, 1966.
11. *Timber Construction Manual*, 1st ed., prepared by the American Institute of Timber Construction, Washington, D.C., John Wiley & Sons, Inc., New York, 1966.
12. *The Encyclopedia of Chemistry*, 2d ed., Reinhold Publishing Corporation, New York, 1966.

13. *Principles and Practices of Heavy Construction,* by Ronald C. Smith, Prentice-Hall, Inc., Englewood Cliffs, N.J., 1967.
14. *Engineering Contracts and Specifications,* 4th ed., by Robert W. Abbett, John Wiley & Sons, Inc., New York, 1963.
15. *Handbook of Industrial Loss Prevention,* 2d ed., prepared by the Factory Mutual System, McGraw-Hill Book Company, New York, 1967.
16. *Organic Coatings: Properties, Selection, and Use,* by Aaron Gene Roberts, Building Research Division, Institute for Applied Technology, National Bureau of Standards, Washington, D.C., 1968.
17. *Roofing: Estimating, Applying, Repairing,* by James McCawley, Shelter Publications, 214 N. Karlov Avenue, Chicago, 1959.

Research Reports

Report Nos. 1 through 15 are available from the Publications Section, Division of Building Research, National Research Council, Ottawa, Canada.

1. "Built-up Roofing," by M. C. Baker, CBD24, December, 1961.
2. "Humidified Buildings," by N. B. Hutcheon, CBD42, June, 1963.
3. "Air Leakage in Buildings," by A. G. Wilson, CBD23, November, 1961.
4. "Moisture Considerations in Roof Design," by G. O. Handegord, CBD73, January, 1966.
5. "Control of Air Leakage Is Important," by G. K. Garden, CBD72, December, 1965.
6. "Vapor Diffusion and Condensation," by J. K. Latta and R. K. Beach, CBD57, September, 1964.
7. "Thermal Considerations in Roof Design," by G. K. Garden, CBD70, October, 1965.
8. "Temperature Gradients Through Building Envelopes," by J. K. Latta and G. K. Garden, CBD36, December, 1962.
9. "Extreme Temperatures at the Outer Surfaces of Buildings," by D. G. Stephenson, CBD47, November, 1963.
10. "Mineral Aggregate Roof Surfacing," by D. C. Tibbetts and M. C. Baker, CBD65, May, 1965.
11. "Fundamentals of Roof Design," by G. K. Garden, CBD67, July, 1965.
12. "Properties of Bituminous Membranes," by P. M. Jones and G. K. Garden, CBD74, February, 1966.
13. "Flashings for Membrane Roofing," by M. C. Baker, CBD69, September, 1965.
14. "New Roofing Systems," by M. C. Baker, CBD49, January, 1964.
15. "Wind Pressures and Suctions on Roofs," by W. A. Dagleish and W. R. Schriever, CBD68, August, 1965.
16. "An Investigation into the Causes of BUR Failures," by Donald E. Brotherson, Research Report 61-2, University of Illinois Small Homes Council, Building Research Council, October, 1961, $1.50.
17. "Roof Failures: Understand Them and Avoid Them," by C. E. Lund, Shelter Publications, 205 W. Monroe St. Chicago, January, 1966.
18. *"Building Science Seminar on Roof Design Proceedings,"* March 28 to April 1, 1966. Publications Section, National Research Council Division of Building Research, Ottawa.

19. "Built-up Roofs in the State of Washington," by Thorkel M. Haaland and Robert P. Darlington, Bulletin 223, Washington State Institute of Technology, State College of Washington, Pullman, Washington, 1956.
20. "Vapor Condensation in Cold Storage Walls," by T. Kusada and W. W. Dorsey, Institute of Applied Technology, National Bureau of Standards, Washington, D.C. 20234, 1968.
21. "Ridge Formation in Roofing," prepared by E. C. Shuman for the Insulation Board Institute and the Asphalt Roofing Industry Bureau, Pennsylvania State University, 230 Hammond Building, University Park, Pennsylvania, 1960.
22. "Premature Failure of Built-up Roofing," by Frank A. Joy, Building Research, The Pennsylvania State University College of Engineering, University Park, Pennsylvania.
23. "Effects of Thermal Shrinkage on Built-up Roofing," by W. C. Cullen, National Bureau of Standards Monograph 89, Washington, D.C., March 4, 1965.
24. "Study of Roof Systems and Constituent Materials & Components," Special Advisory Report No. 6, Building Research Advisory Board, Division of Engineering & Industrial Research, National Academy of Sciences, National Research Council, Washington, D.C. 20418, 1964.
25. "Fire Tests of Building Construction and Materials," UL263, Underwriters' Laboratories, Inc., Northbrook, Ill. 60062, May, 1959.
26. "General Motors Corporation Fire, Livonia, Michigan, August 12, 1953," report by The Factory Insurance Association, Hartford, Connecticut.
27. "Report on Roofing Construction at Dulles International Airport," by W. C. Cullen, Project 10447, Building Research Division, Institute of Applied Technology, National Bureau of Standards, Washington, D.C. 20234.
28. "Refrigerated Storage Installations," Publication No. 759, National Academy of Sciences, National Research Council, Washington, D.C. 20418, 1960.
29. "Moisture Content of Roof Insulation," Department of the Air Force, Memorandum 328CES-EC, 1967.

Codes and Standards

1. "Uniform Standards and Specifications," Structural Wood Fiber Products Association, 1966.
2. "Tentative Recommendation, Built-up Roofing," CSI B12.1-60T, Construction Specifications Institute, 1717 Massachusetts Avenue, Washington, D.C. 20036, 1960.
3. "National Building Code," 1967 ed., American Insurance Association, Engineering and Safety Department, 85 John Street, New York 10038.
4. "Insulated Steel Deck, Acceptable Class I Construction, Loss Prevention Data," Construction I-28S, Factory Mutual Engineering Corporation, February, 1968, Factory Mutual Engineering Corporation & Association, 1151 Boston-Providence Turnpike, Norwood, Mass. 02062.
5. "Building Materials List," Underwriters' Laboratories, Inc., January, 1968. 333 Pfingsten Rd., Northbrook, Ill. 60062.
6. "Test Method for Fire Hazard Classification of Building Materials," 3d ed., UL723, Underwriters' Laboratories, Inc., 333 Pfingsten Rd., Northbrook, Ill. 60062, August, 1960.
7. "Test Methods for Fire Resistance of Roof Covering Materials," 1st ed., UL790, Underwriters' Laboratories, Inc., 333 Pfingsten Rd., Northbrook, Ill. 60062, September, 1958.

8. "Bulletin of Research for Determining Wind-Uplift Resistance of Roof Assemblies," Underwriters' Laboratories, Inc., 333 Pfingsten Rd., Northbrook, Ill. 60062, April, 1962.
9. "Roof Maintenance Manual," by H. R. Snoke & W. C. Cullen, National Bureau of Standards Report 3994, 1955, Building Research Division, Institute of Applied Technology, National Bureau of Standards, Washington, D.C. 20234.
10. "Basic Building Code," 1965 ed., Building Officials Conference of America, Inc., 1313 East 60th Street, Chicago 60637.
11. "Building Code Requirements for Minimum Design Loads in Buildings and Other Structures," American National Standards Institute, 1430 Broadway, New York, N.Y. 10018.

Articles

1. "Recommended Practice for Insulated Built-up Roofs," by C. E. Lund, *AIA Journal*, April, 1965.
2. "Research Looks into Roofing Failures," by Foster C. Wilson and Miles E. Jacoby, *Architectural Record*, October, 1962.
3. "Building Movement Can Damage Built-up Roofing Systems," by Werner H. Gumpertz, *Architectural Record*, September, 1966.
4. "New Theory for What's Behind Built-up Roofing Failures," by Kenneth Tator, *Architectural Record*, November, 1966.
5. "Steel Roof Collapse Injures 15," *Engineering News-Record*, August 3, 1967; "Plugged Drains Blamed for Steel Roof Collapse," *Engineering News-Record*, August 17, 1967.
6. "Roof Deflection Caused by Rainwater Pools," by Robert W. Hanssler, *Civil Engineering*, October, 1962.
7. "Specifying Building Insulation," by Abraham I. Tenzer, *Architectural & Engineering News*, February, 1963.
8. "Roof Failures: Whose Responsibility?", by Edward T. Schreiber, *Architectural & Engineering News*, November, 1964.
9. "Insulation: Ally or Enemy," by Miles E. Jacoby, *Specifier*, August, 1967, Construction Specifications Institute, 1717 Massachusetts Ave., N.W., Washington, D.C. 20036.
10. "Case for the 'Systems' Specification," by Frank L. Couch, CSI, *Specifier*, August, 1967, Construction Specifications Institute, 1717 Massachusetts Ave., N.W., Washington, D.C. 20036.
11. "Effect of Insulation on the Weathering of Smooth-Surfaced Built-up Roofs Exposed to Solar Heating," by W. C. Cullen and W. H. Appleton, *American Roofer*, March and April, 1963.
12. "Designing Low Slope Roof Systems," by Robert M. Stafford, CSI, *Specifier*, May, 1968, Construction Specifications Institute, 1717 Massachusetts Ave., N.W., Washington, D.C. 20036.
13. "Blowoffs, Cause and Cure, and Cause for Concern," by James McCawley, *American Roofer*, April, 1968.
14. "Blowoffs Concern You!", *American Roofer*, July, 1968.
15. "Working Correctly with Elastomer Roofing," by Stanley Schmitt, *Architectural & Engineering News*, July, 1965.
16. "Elastomeric Roofing," by Werner H. Gumpertz, *Architectural & Engineering News*, September, 1969.

Index